The Lunar Exploration Scrapbook

A pictorial history of Lunar Vehicles
by Robert Godwin

This book was completed on Thursday July 27th 2007 and is respectfully dedicated to Eric Dean Blackwell, Todd Ivens and Charles Glen May—three pioneers on the long, high road.

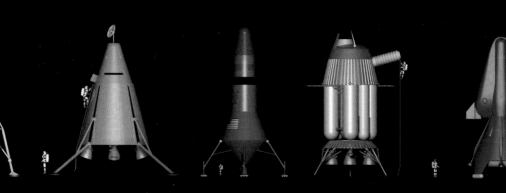

Acknowledgments:
Thanks to my nephew Geoff Godwin for the crash course in 3D modeling and to my daughter Emily for slaving over the photocopier. Also to the following people for research assistance and insights: Mark Kahn at NASM, Nadine Andreassen. Colin Fries, Steve Garber, John Hargenrader and Jane Odom at NASA HQ. Mike Gentry & Benny Cheney at JSC, Irene Willhite, Frederick Ordway III, Gordon Woodcock and Dennis Wingo in Huntsville, Don Beattie, Dr Hermann Koelle, Ron Miller, Dr Harrison Schmitt, Neil Armstrong, Dr David Stephenson and to my ubiquitous astronaut avatar who got screened in over so many models, we'll just call him Matt...

All rights reserved under article two of the Berne Copyright Convention (1971).
We acknowledge the financial support of the Government of Canada through the Book Publishing Industry Development Program for our publishing activities.

Published by Apogee Books, Box 62034, Burlington,
Ontario, Canada, L7R 4K2, http://www.apogeebooks.com
Tel: 905 637 5737

Printed and bound in Canada
The Lunar Exploration Scrapbook by Robert Godwin
©2007 Robert Godwin
All imagery of 3D renderings ©2007 Robert Godwin
Photos and diagrams courtesy NASA and USGS
ISBN 978-1-894959-69-8

Introduction

The Lunar Exploration Scrapbook began as an attempt to organize the data piling up in the Apogee filing cabinets. It soon became evident that we had a wealth of rarely seen designs accumulating in our files, and many of them had never existed beyond basic blue-prints or line-art diagrams. Since the birth of the space age there have almost certainly been thousands of unrealized designs for lunar vehicles and this book undoubtedly barely scratches the surface. However, a comprehensive register would probably be an impossible undertaking, since so many of those ideas have been lost or buried, and so this book will stick to some of the more serious designs developed by the major contractors up until the end of the Apollo moon landings. Some of the data behind these designs is still classified and so it was not possible to provide a consistent range of information for each and every design. With that in mind it seemed to be more practical to present the information as a scrapbook of bits and pieces.

In the following pages you will find three basic types of vehicle. Lunar landers, lunar roving vehicles and lunar flying vehicles. In amongst the rovers you will find vehicles that also served as mobile laboratories such as the MOLAB, while in amongst the landers you will find vehicles that border on the verge of moon base habitats, such as the SHELAB. Habitats have been restricted to those that were proposed using the LLV or modified Lunar Modules; another entire book could be done on moon bases.

In the instances where hardware was actually built, some effort has been made to include a real photograph of the vehicle. In other instances artist impressions from the NASA archives appear instead. Finally, and obviously, the bulk of the book is taken up by computer renderings based on the original line-art drawings. Each rendering is shown from at least two sides and then, as it might have looked in a hypothetical in-situ.

The reader may well notice that some of the rendered designs are lacking fundamental equipment such as egress hatches, reaction thrusters or ladders. In these instances it was not possible to find complete drawings to be able to determine their proposed location. Rather than resort to speculation it was decided not to add these missing features rather than risk putting them in the wrong place. Consequently some of the renderings look a little sparse. Sometimes the original contractor or NASA drawings showed a major design feature from only one angle, and so rather than preclude that vehicle entirely, a little speculation did take over in an attempt to create something that made sense based on that one angle view, symmetry and launch-friendly aerodynamics.

This book is entirely subjective, based on things that have sparked the author's interest. Any reader looking for a comprehensive objectivity is likely to be disappointed. Some of the designs within came from literally the last existing copy of a document. Although this might be perceived as a history book I have made no attempt to draw any modern conclusions, it is simply a fanciful attempt at presenting many extraordinary, and often extremely rare ideas, in one place.

Robert Godwin (Burlington 2007)

[BIS 1938-1949]

In 1938 the first serious efforts to design a lunar spacecraft were made by a multi-disciplinary committee of the prestigious *British Interplanetary Society*. The committee included young astronomer Arthur C. Clarke and engineer H.E. Ross, who would also be one of the first to propose a viable method for actually performing a moon flight. In an article written in 1969 Ross would credit most of the BIS design to turbine engineer, R.A. Smith and committee chair, Happian Edwards. It should be noted that Smith was also an excellent artist. The original drawings which appeared in January 1939 showed that all but one of the banks of solid rockets would be discarded during final approach clearing the way for six banks of liquid rockets to provide fine landing control. The final central bank of solids would be used for the ascent.

The 1938 BIS lunar lander was to have been manned by a crew of three and would have been spun around its Y axis at a rate of one revolution every three seconds. This rotation was provided to create a simulation of one gravity. A catwalk was placed around the inside wall of the cabin for the crew to make their way around the ship. Each crewman would lie flat on their backs on axially oriented couches. This extraordinary vehicle was to have been launched on five stages, each one equipped with 169 rocket motors. The landing craft would have left the moon atop hundreds of small solid rockets. The BIS lunar spacecraft would have flown straight to the moon, landing the entire spacecraft atop six hydraulic shock absorber legs. These legs would be stored inside the aerodynamic fairing of the vehicle until needed for the actual landing. Re-entry was dismissed as something that would take place at "low velocities".

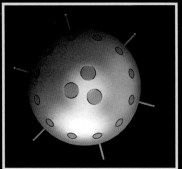

The 1938 BIS lunar lander was to be launched aboard literally hundreds of solid fuel rockets. It would land on the moon using a combination of solid and liquid motors. The lander would have stood about 11 feet tall.

1939 artist rendering of vehicle in flight (bottom left)

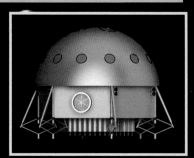

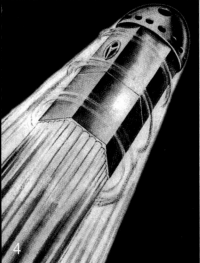

The 1949 BIS lunar lander

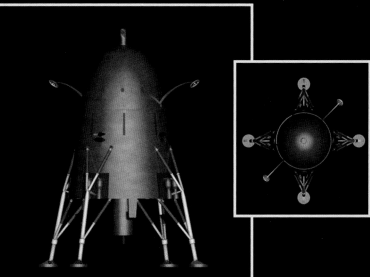

The importance of the BIS lunar spacecraft should not be underestimated. It was the first time anyone, with real experience and technical skills, had actually tried to design a lunar lander that would work; based on what we really understood about the moon. Just a few short months after the BIS committee published their findings Europe was thrust into World War II and so for the next six years very little discussion took place on the subject of lunar travel.

Around 1949-50 R.A. Smith rendered a substantially revised variant of the BIS lander, which unlike its predecessor was brilliantly practical, and bore many striking similarities to some of the designs you will encounter in the following pages. Arthur C. Clarke provided an in-depth description of Smith's new BIS system in his 1951 book *Exploration of Space*. The lander was to have a descent stage that doubled as a launch pad; a lunar orbit rendezvous technique was to be used for refueling; a radar altimeter for lunar approach; and the vehicle itself was a spidery four legged structure with paper thin walls. Smith and Clarke worked together again in 1954 on a book called *Exploration of the Moon*. Smith's art again showed the spacecraft in several different settings, but it was fundamentally unchanged from the earlier work. What is perhaps of key interest is just how much this design seems to have influenced what actually flew in 1969. Many other larger and wildly optimistic schemes appeared in the intervening decades, but the fundamental concept of a light-bodied, four-legged lander that used its lower half as a launch platform seems to have weathered all of these later ambitious permutations.

[Von Braun Lunar Lander 1952]

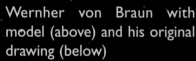

Wernher von Braun with model (above) and his original drawing (below)

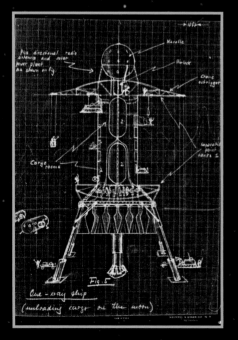

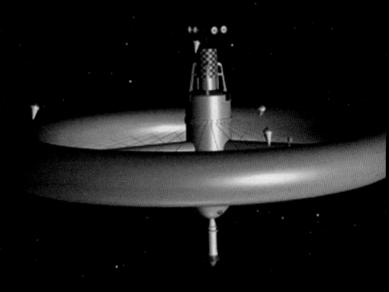

NASA's Apollo program can trace its heritage back to the science fiction of the 19th century but it would not be until the conclusion of World War II that America would take the tenuous first steps towards an actual lunar program. Driving this campaign, from inside an internment camp in New Mexico, was the German rocket engineer, Wernher von Braun. Even before the war von Braun had been working on rocket systems capable of reaching space, but once he became a prisoner in the United States he found himself free to speculate in a way that had been impossible in Nazi Germany. His ideas for a lunar lander were outrageously bold and larger than life. At that time his imagined method for landing a spacecraft on the moon required a massive construction effort in low Earth orbit, before dispatching a titanic landing craft on its way to the moon. This early von Braun moon-ship appeared in *Collier's* magazine in 1952 and again in the 1953 Viking book, *Conquest of the Moon*, beautifully illustrated by noted space artist Chesley Bonestell. Von Braun envisioned it in two different versions, as a passenger ship and modified as a cargo ship.

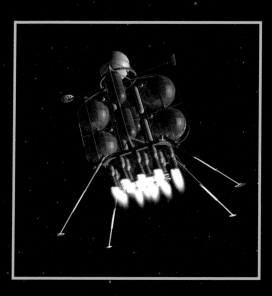

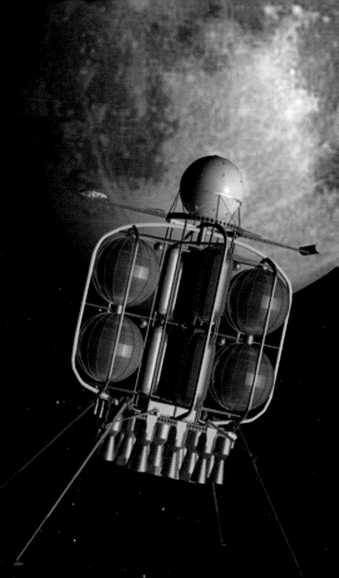

The manned version of Von Braun's 1952 lander is seen here. The large round tanks would be jettisoned during the mission.

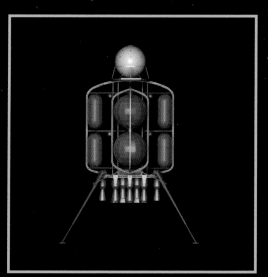

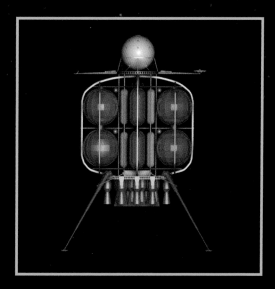

[Project Horizon 1959]

In 1957 the space race began with the launching of Sputnik and the American military shifted their efforts into high gear. NASA would be established in 1958 but both the US Air Force and the Army continued to put some considerable effort into creating lunar landing architectures right into the first few months of 1959. Not surprisingly both military propositions were large, expensive and ambitious. Very few books even mention these two programs. Even though they are not from NASA, one of them was intimately connected with von Braun, so both have been included here for comparison purposes (see LUNEX on p 32). On December 15th 1958 rocket engineers Wernher von Braun, Ernst Stuhlinger and Heinz-Hermann Koelle traveled to Washington to make their case for sending humans to the moon by seeking continued financial support for the Juno V (which was to become the proposed Saturn I booster).

Horizon in lunar simulation chamber

The von Braun and JPL teams, still part of the Army at this time, then met with NASA on February 5th 1959. The subjects discussed included circumlunar vehicles, hard lunar impacts, lunar satellites and soft lunar landings. Their preliminary findings were issued on May 1st 1959. This US Army plan became known as *Project Horizon* and would be led by Koelle.

The Von Braun/Army lunar landing system would require a truly enormous booster that stood 438 feet tall, harnessing 12 million pounds of thrust on its first stage. *Horizon* would have more than one configuration, allowing for both orbital payloads (for supplying space stations) as well as a lunar landing component. The command module (or re-entry module) would be modeled after the recently perfected *Juno* warhead and would sit nestled inside the lunar lander. The orbital module was to have carried up to 16 crew, while the lunar lander would stand about 60 feet high and take at least two people to the lunar surface. Since the massive booster required for this was unlikely to be built, von Braun drew up an alternative version which would have required the launching of six of the new proposed Saturn C-3 boosters. Five of these launches would take the necessary fuel into low Earth orbit and they would then cluster around the sixth vehicle (which would have the lunar lander attached) to supply it with fuel for the lunar trip. The *Horizon* proposal was not fully completed until February of 1960 but by that time NASA had already taken over most of the country's space activity and had requested that von Braun's team consider using

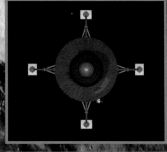

their proposed Saturn rocket, and so the monumentally optimistic Horizon system never went beyond the four volumes of paperwork generated in Huntsville.

In July 1959 NASA Chief of Propulsion Development, Milton Rosen, and future nuclear and laser propulsion expert, Carl Schwenk, started to draft their own ideas for how a lunar architecture might look. Rosen and Schwenk had been inspired by the engine contract granted to Rocketdyne a few weeks earlier. This contract called for the construction of the biggest rocket engine ever made, the 1.5 million-pound thrust F-1.

Wernher von Braun with Horizon behind him (above)

Original Horizon drawing (left)

12 million pound booster for Horizon (below left)

Cutaway showing second stage and crew compartment of Horizon lander

[Rosen-Schwenk/NOVA 1959

At the same time that *Horizon* was brewing in von Braun's office, the Rosen/Schwenk team inaugurated their super-booster design dubbed *Nova*, which they first revealed publicly on 31st August 1959 at the *Tenth International Astronautical Congress* in London. Built on top of an array of F-1 rocket engines, it was believed that *Nova* could be built on a scale large enough to send a lunar landing system directly to the moon, exactly like *Horizon*. This method of flight was named "Lunar Direct" since the spacecraft would literally be built to fly from the launch pad straight to the moon, with no interludes in orbit.

In May of 1961, immediately after the flight of the first American in space, President John Kennedy stepped up the pace of the campaign for a lunar landing. His announcement—to accomplish the seemingly impossible within nine years—would catalyze the process and force America's engineers to rethink their previously grandiose plans. With less than a decade to fulfill their President's directive, NASA's engineers would have to forego the massive architectural dreams of the 1950's and come up with something that was actually feasible. Initially, attention returned to Rosen's gigantic, hulking *Nova* rocket system. *Nova* would not be quite as tall as *Saturn*, at least not in its first iteration. The 1959 *Nova* would stand 220 feet high, 48 feet in diameter and would weigh about 3000 tons. It would be propelled by six F-1 engines. It had four stages to get the lander to the moon and then the lander's ascent module would constitute a fifth earth-return stage. The lander would weigh about 18 tons.

Just four weeks before President Kennedy's announcement, Koelle had written down the specifications for an upgrade to the *Nova* system that featured eight F-1s on the first stage, two F-1's on the second stage, four of the proposed hydrogen powered J-2's for the third stage, six Pratt and Whitney RL-10s on the fourth stage and finally two more RL-10s on the fifth stage. Later both solid rockets and nuclear rockets would be considered for *Nova*, in an attempt to squeeze even more out of the colossal design.

Rosen and Schwenk would recognize the need for high energy propellants on the upper stages of a moon rocket and they also understood that using the proposed Saturn (in its earliest configuration) might require up to 24 launches to accomplish a satisfactory lunar exploration program. They therefore proposed a direct method using what they referred to as an upgraded Mercury spacecraft. It was to be 14 feet high with two floors. The lower floor contained couches, controls, communications and a folding airlock while the top floor contained food, power supply, EVA gear and work space. The outside of the capsule was coated with an ablative material to accomodate reentry.

NOVA contractor model (above) and a later iteration of Nova compared to Saturn (left) Nova co-creator Milton Rosen (below)

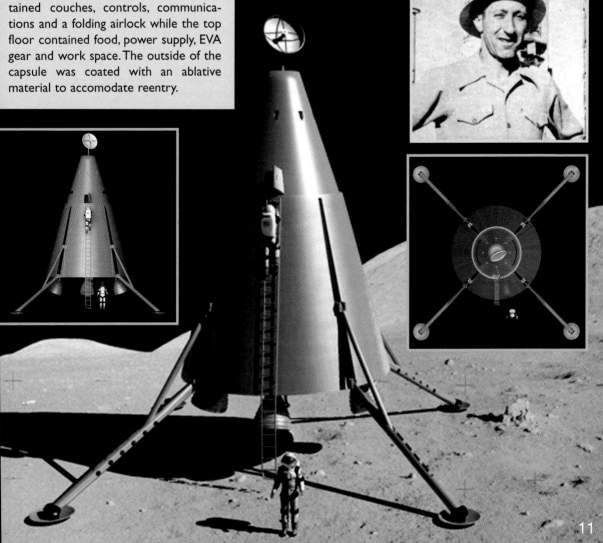

LUNAR RETURN MISSION
LAUNCHING

TAKE-OFF FROM MOON WITH FIFTH STAGE

M-1 ENGINE POTENTIAL APPLICATIONS
UPPER STAGES
FUTURE LAUNCH VEHICLES

FIRST STAGE
FUTURE LAUNCH VEHICLES
(REQUIRES MODIFICATIONS)

LUNAR RETURN MISSION
SECOND STAGE FIRING

JETTISON FIFTH STAGE

LUNAR RETURN MISSION
THIRD STAGE FIRING

CAPSULE REENTERS

For a short time in the early sixties Rosen and Schwenk's 1959 slides (left) became the de facto model for a lunar landing plan. One company that jumped on that band wagon was *Aerojet*, a Sacramento-based supplier of space and missile hardware. In fact *Aerojet* had been the driving force behind the giant 2 million pound-thrust *M-1* engine, also proposed for *Nova* (above). Even a single stage to orbit concept was considered with the M-1 but was soon dropped.

FOURTH STAGE ROTATION

RETURN TO EARTH AND RECOVERY

DESCENT TO LUNAR SURFACE

LUNAR LANDING

EXPLORING THE MOON

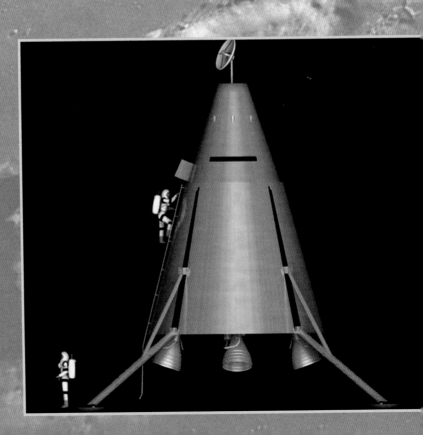

12

[Aerojet Moonmobile 1960]

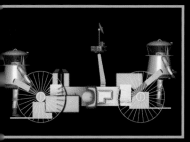

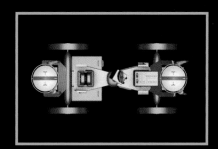

The Aerojet Moonmobile carried life support for two exploring astronauts in hard shell suits and was powered by fuel cells. It was to have been capable of traversing over 1000 miles at a top speed of 5 mph.

The *Aerojet* designers snapped up Rosen's slides and produced their own presentation for a complete lunar base that would be manned by up to 12 crew and would include roving vehicles, like the Mk1 Moonmobile, as well as supply ships and a very conspicuous hard-shell lunar space suit.

[Moonsuits 1958-1971]

Aerojet Moonsuit

The Aerojet moon suit was designed by JPL engineer Allyn B. "Hap" Hazard and would be supplied with its consumables from the accompanying rover. It would not only allow the crewman to pull his arms inside, to accomplish the coveted nose scratch, but also to cook food inside the suit or simply roll onto his back and take a nap! The *Aerojet* suit became famous for having appeared on the cover of LIFE magazine in April of 1962 (where it was credited to *Space-General Corporation*, the new name for Aerojet and SEC combined) before being truly immortalized when it was co-opted by the *Mattel* corporation for its, now legendary, *Major Matt Mason* toy line in 1966. Of course the other moon suit seen at right, with the black rubber joints also became famous through the toy makers efforts. This design came from NASA spacesuit engineer Leonard Shepard. The "constant volume" rubber joints were deemed essential if an astronaut were to be able to move against the severe pressure differential between the suit and lunar vacuum. It was designed in 1958 and would be continually improved before Shepard patented the Apollo lunar suit ten years later (opposite top).

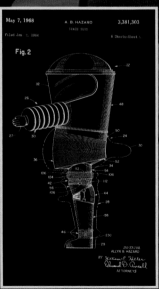

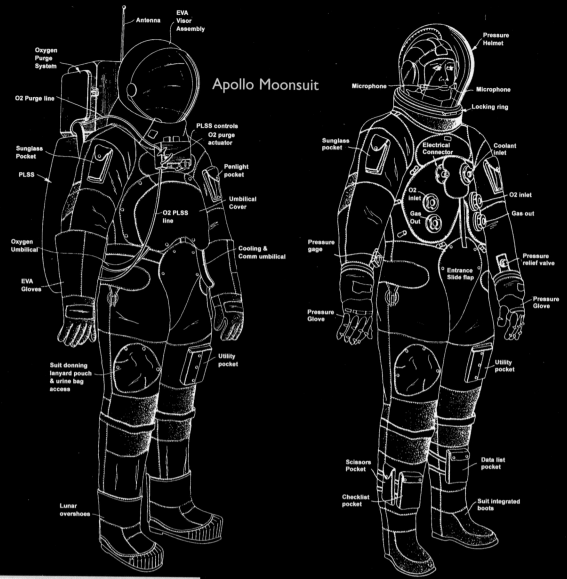

Apollo Moonsuit

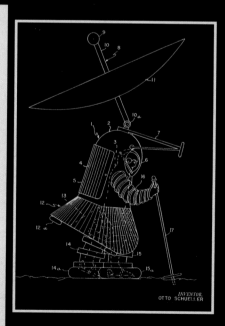

Another variation of hard-shell suit that appeared a little later was that from Republic Aviation (left). It bore some minimal resemblance to the Aerojet suit but looked even more bizarre, and even a little antiquated. It was, however, nowhere near as peculiar as the 1961 creation by Air Force designer Otto Schueller (right) in which the astronaut could sit down in the suit and was protected by a huge overhead radiation umbrella.

[Lunar Direct 1961]

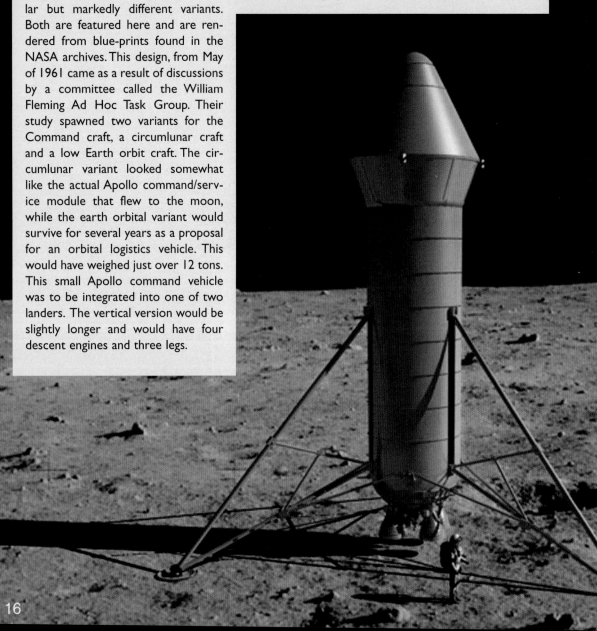

While different versions of NOVA came and went across drafting tables at Langley and in Huntsville, another team of NASA engineers were trying to nail down the final design for a lunar descent engine. This inevitably led to discussions about how such a vehicle would approach the moon. Most traditional ideas assumed it would back its way down, on a column of exhaust (exactly like the science fiction shows of the time), but another idea was to fly it in like an aircraft and land horizontally. The only difference being that this almost horizontal approach would have still flown in tail-first. This led to the release of a series of famous NASA graphics that showed Apollo in two similar but markedly different variants. Both are featured here and are rendered from blue-prints found in the NASA archives. This design, from May of 1961 came as a result of discussions by a committee called the William Fleming Ad Hoc Task Group. Their study spawned two variants for the Command craft, a circumlunar craft and a low Earth orbit craft. The circumlunar variant looked somewhat like the actual Apollo command/service module that flew to the moon, while the earth orbital variant would survive for several years as a proposal for an orbital logistics vehicle. This would have weighed just over 12 tons. This small Apollo command vehicle was to be integrated into one of two landers. The vertical version would be slightly longer and would have four descent engines and three legs.

Lunar Direct Model, note the flat plates on the legs, not apparent on the blue prints below.

Lunar Direct Blueprint

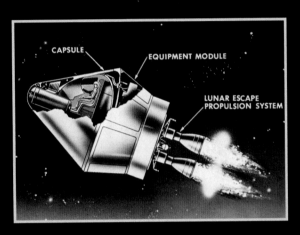

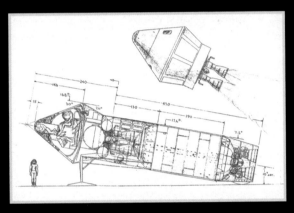

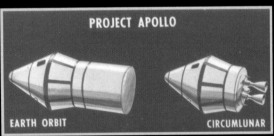

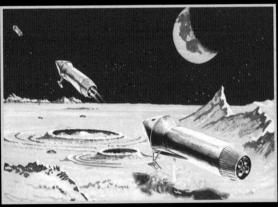

Apollo Command ship variants to be used for *Lunar Direct* (above) these would have weighed about 12.5 tons

Lunar Direct Blueprint (top right) the lander would have weighed 75 tons.

Artist impressions showing landing and take off of *Direct* sidelander (right)

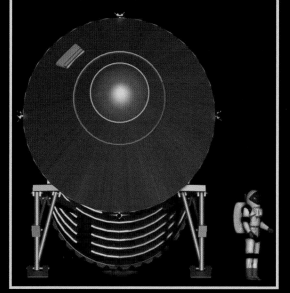

The version of Fleming's Lunar Direct that would land horizontally would be almost 60 feet long and featured four main "retro" engines accompanied by a central vertically directed stabilizing "descent" engine. It would have used skids on the front and a disc shaped landing pad on the bottom of the fuselage. Fleming's committee estimated that this version of Apollo would land in late 1967 and would cost almost $12 billion.

One of many proposals for Nova to carry Apollo Lunar Direct (right)

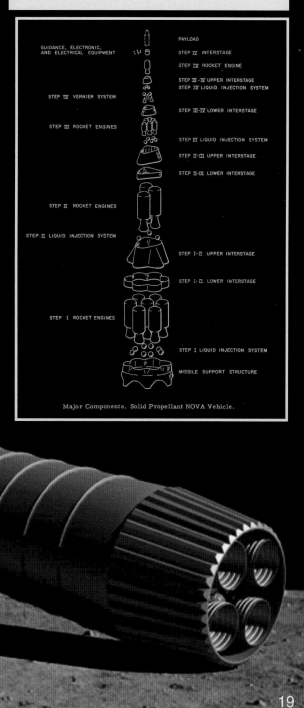

[Saturn vs Nova]

Between its first inception, in the summer of 1959, and its final cancellation in 1964, the *Nova* would at one point evolve to include a staggering eighteen F-1 engines and 3 of the even larger M-1 engines, for a cumulative total of 32.4 million pounds of thrust! But *Nova* would be destined to exist only on the drawing boards, and it would go through many permutations before it was realized that it was easier and cheaper to use the smaller *Saturn*. Had Apollo flown on *Nova* the *entire* lunar vehicle would have landed on the moon, like something from a 1950's science fiction movie, tail first. There would have been no refueling stops or orbital rendezvous.

The designers across America came up with a variety of spacecraft to fly atop *Nova*, using the proposed "lunar direct" method. One of the earliest problems that had to be resolved was the shape and structure of the spacecraft. At least one committee was shown a flying saucer design for the command module, created by engineer Alan Kehlet (co-inventor of the Mercury spacecraft). Although it would never fly (as far as we know!) it was patented in 1963 and it would be taken seriously enough to pop up as part of the *Martin Corporation's* proposals for *Apollo (see page 29)*. Almost all of these designs were based on a direct flight to the moon, which would mean using a huge booster carrying massive quantities of fuel. The considerable expense of developing the giant *Nova*, and then the immediacy of Kennedy's challenge, made it less and less likely that *Nova* would see the light of day; so in August 1961, another NASA committee, chaired by Col. D.H. Heaton, turned thoughts back to von Braun's preferred method—one which would require some construction in low Earth orbit.

This second method was called Earth Orbit Rendezvous (EOR) and it would require several launches of the much smaller *Saturn* rocket. EOR would place the parts for the lunar spacecraft into Earth orbit where it would be met by tankers of fuel. This method had been one of the alternatives for *Project Horizon* where it was thought that up to six launches of the proposed *Saturn C-3* would be needed to bring all of the necessary parts together. Heaton's committee reduced this to four launches. The arrival of the *C-5 Saturn* brought the number down to two launches. Once all of the components connected in Earth orbit, the ship would then continue on to make a lunar landing. The problem with this method was that there were still at least one too many launches. It was 13% cheaper than *Lunar Direct* but it was still expensive and risky. If one launch failed, it could mean the end of the entire mission. However, this method still found favor, since it could be achieved with a smaller, and more feasible booster than *Nova*, and also because most of the hazardous maneuvers took place closer to the Earth. This had some considerable appeal as it was much safer for the crew to be only minutes from the Earth if anything went wrong.

The *Saturn* booster system had been under development as the Juno V at *Redstone Arsenal* in Huntsville, Alabama, since September of 1958. Once again the extremely powerful F-1 would be the proposed engine to power the booster. On 16th December 1959 the Army allowed NASA to establish a permanent presence at *Redstone Arsenal*. The final control of the *Saturn* project was handed over to NASA, by direct order of the President, in January of 1960.

An early artist's painting and model of the Saturn proposed by Von Braun, Koelle and Stuhlinger to support a US Army lunar program.

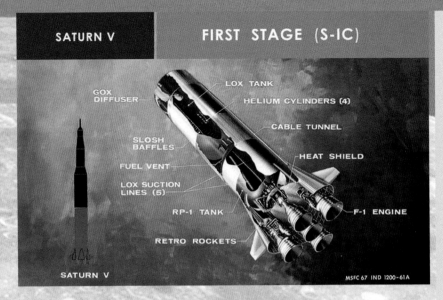

NASA diagrams showing the three main stages for the Saturn V launch vehicle.

The SIC (at top) was powered by five Rocketdyne F-1 engines generating 7.5 million pounds of thrust at sea level and almost 9 million pounds at burnout.

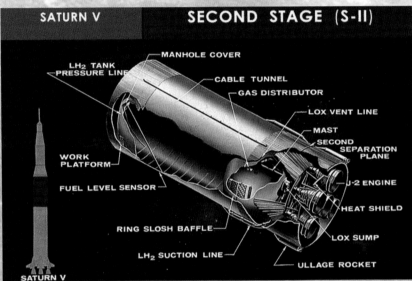

The S-II was the most powerful hydrogen powered rocket in the world with five J-2 engines.

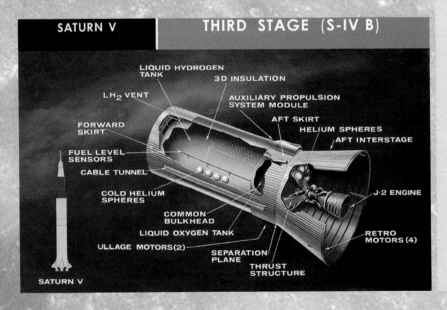

Finally, the S-IVB at bottom left was powered by a single J-2 engine with a restart capability. This engine could power the Apollo both into and out of Earth orbit.

1963 NASA comparison chart, showing engines proposed for the Apollo program (bottom)

Kennedy Space Center during the heyday of Apollo, with three mobile crawlers outside the Vehicle Assembly Building (left)

Saturn V leaving the Vehicle Assembly Building in 1969 (centre)

[US Army/JPL Prospector 1959-1960]

Before the Apollo program really got up to speed an earnest effort was made by the Jet Propulsion Laboratory to show that robots could do much of the ground work, and possibly all of the work that an astronaut could do. This *Prospector* program was on obvious extension of the upcoming Ranger and Surveyor programs and would use a Saturn C1 booster to place up to 15,000 pounds on the moon. Early plans for this mission came from the *Project Horizon* studies in 1959. Prospector would have the capability of delivering a large mobile laboratory to the moon very similar in principle to the Martian rovers of later years. JPL/NASA turned to Goodyear and the Arnold Schwinn bicycle company to develop large wheels for the Prospector rover. This very large wheel was inflatable and rubberized but could not withstand the lunar cold so a more complex wire wheel was developed. The wire wheel research was then used by Bendix, GM and Boeing for their later MOLAB studies. The overall approach for Prospector, using a landing stage and a retro stage to deliver an assortment of payloads represented a very early prototype of the paradigm later adopted by most of the contractors for the Apollo Applications Program studies using the LM.

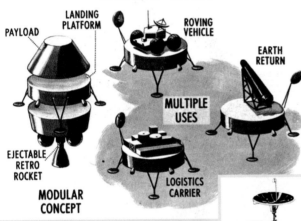

The basic premise for the *Prospector* rover was to land and search for appropriate manned landing sites as well as provide transport to and from manned bases. This same methodology was later adopted by the Soviets for their Lunokhod program.

[Lunar Surface Rendezvous 1961]

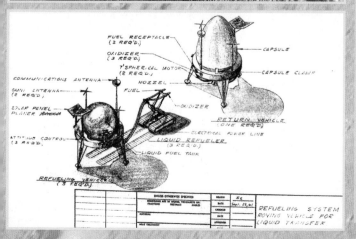

An announcement was made to the contracting community at NASA's *Langley Research Center* on September 13th 1960 and an official RFP (request for proposals) for a lunar spacecraft was issued. Needless to say the entire contracting community scurried away to see what they could devise in an effort to win the incredibly lucrative contract. Within less than a month fourteen of the eighty-eight attending contractors submitted bids, these included *Boeing; Convair/Avco; Cornell/Bell/Raytheon; Douglas; General Electric/Bell; Goodyear; Grumman/ITT; Guardite; Lockheed; McDonnell; Martin; North American; Republic; and Vought*. The following May the design proposals started to arrive at Langley.

Another option that was briefly considered was that of lunar *surface* rendezvous. This idea originated at JPL, during the discussions in 1959, presumably because the JPL team were focussing on unmanned and robotic equipment. The idea was to take advantage of the *Surveyor/Prospector* technology and have several unmanned refueling vehicles and an Earth return vehicle, land on the moon in advance of the arrival of the manned lander. The NASA Marshall center under Koelle did not study this method but JPL issued a request the following September. At least one company, *Space Craft Inc* of Huntsville, produced an elaborate presentation that included extensive blue-print designs and photography of carefully crafted models. Several of these images can be seen here. Lunar surface rendezvous was dismissed almost as soon as it arrived because it had few advantages, either technically or fiscally, because it required six Saturn launches.

[GD/Convair M1 1961]

On May 15th 1961, just ten days after America's first manned space flight, on the exact day of President Kennedy's exhortation to reach the moon, *General Dynamics/Convair* presented a three module system that could be upgraded to allow both orbital and lunar landing missions. It came with a price tag of over a billion dollars. They also came up with a proposal for a lenticular re-entry vehicle which would have been launched inside a Juno style fairing. There seems to be no evidence that Convair adapted the lenticular shape for a lunar landing craft.

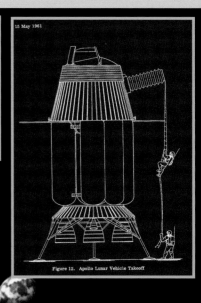

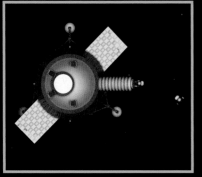 

Convair contractor model (below) The unusual shape could be launched on a *Saturn C-2* or directly using *Nova*.

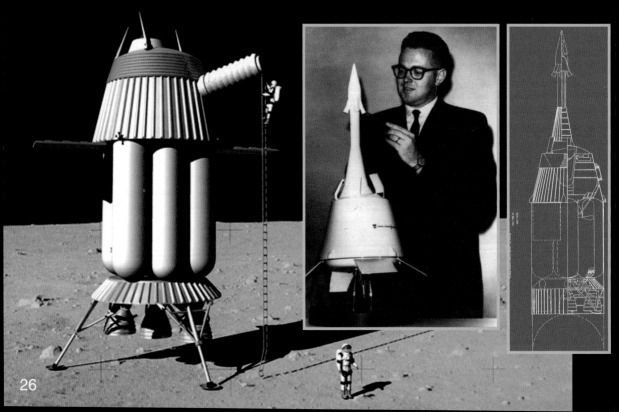

[General Electric-D2 1961]

James Webb holds a GE D-2 reentry module contractor model. It would also have been launched by the Saturn C-2 (left) Inside the D-2 (below)

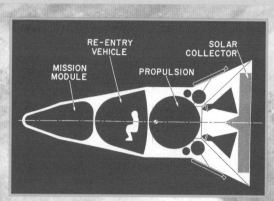

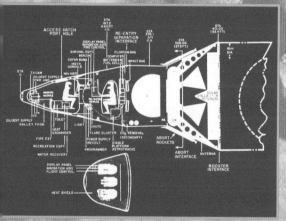

General Electric also presented several designs for the main Apollo spacecraft. The D-2 included a fairly typical blunt-bodied reentry vehicle placed inside a launch-friendly fairing. The Soviet Soyuz lunar vehicle looked suspiciously similar to the GE-D2, with its forward mission module, central reentry module and rear propulsion module. The GE-D2 does not appear to have made any accommodations for a landing, but it did give the crew a much larger living area for a considerably smaller launch mass. GE also looked at a lenticular reentry vehicle and various winged gliders. An agreement was made between NASA and GE in February of 1962. GE was given a multi-year $635 million contract to be responsible for all integration and checking of the Apollo vehicle. In exchange for this they agreed to not seek any further Apollo contracts, since they would get to see all of the competition's designs. By all accounts the GE-D2 was a far superior design to what ultimately flew as the Apollo CSM. Engineers at Langley and Ames had spent years trying to determine the best shape for reentry. A bewildering array of tested shapes were then utilized by contractors like GE and GD/Convair in their bids for Apollo. Some of the ID numbers for these shapes were assigned by NASA while some were contractor numbers.

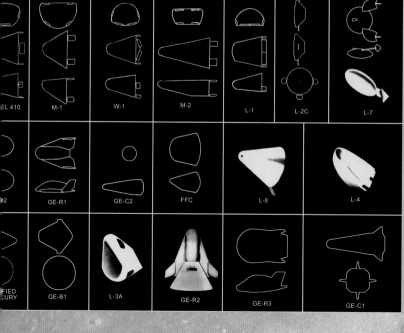

[Lenticular Apollo 1960-1961]

Three weeks after GD/Convair and GE made their grab for the gold ring, the *Martin Company* presented a staggering 9000 page study with their own solutions to lunar flight. Their proposal featured at least three different configurations for command modules. One would be the lenticular or flying saucer shape known as *L-7*, one would be like a blunted lifting-body shape and the other would look remarkably like the *Rockwell Apollo* command module that actually flew to the moon several years later, they called this one the *Modified Mercury* or *L-20*. The lenticular shape and various derivatives were tested extensively at Langley and by Bell and GE, but were found to have problems in high wind scenarios during landing.

Martin also produced at least three totally different designs for a lunar descent stage. Surely the most bizarre of these three designs involved a type of inflatable landing gear. Although the three command modules and three descent modules may not have been designed as specifically matching sets, in the accompanying renderings you will see the flying saucer paired with the three-leg descent stage and the stubby lifting body with the inflatable-legged descent stage.

Martin was NASA's first choice to build Apollo and the company management were so convinced that they had won the contract they announced it over the PA in their factories. However, NASA management decided at the last minute to favor *North American Aviation* because the aircraft manufacturer already had a long working history with both NASA and its historical predecessor the NACA. At the time that *North American* won the contract no one yet knew how the lunar mission was to be achieved. *Lunar direct* or *Earth orbit rendezvous* were the only serious contenders and so the early design work on Apollo made the assumption that the whole vehicle was going to land on the moon.

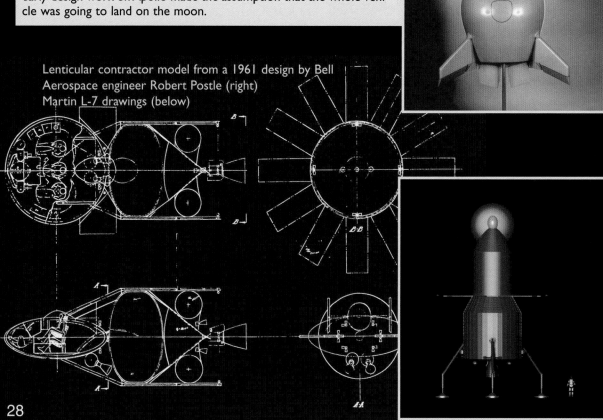

Lenticular contractor model from a 1961 design by Bell Aerospace engineer Robert Postle (right)
Martin L-7 drawings (below)

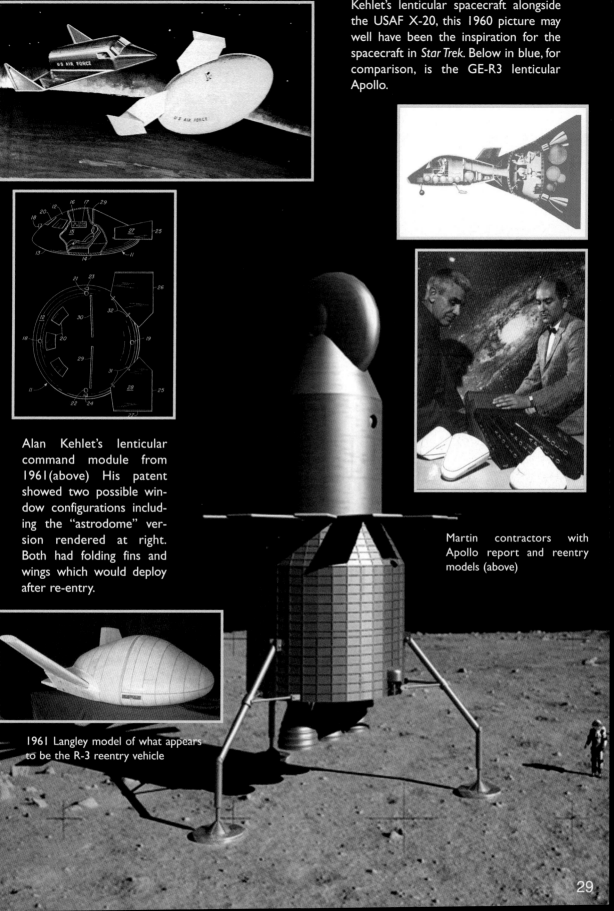

Kehlet's lenticular spacecraft alongside the USAF X-20, this 1960 picture may well have been the inspiration for the spacecraft in *Star Trek*. Below in blue, for comparison, is the GE-R3 lenticular Apollo.

Alan Kehlet's lenticular command module from 1961 (above) His patent showed two possible window configurations including the "astrodome" version rendered at right. Both had folding fins and wings which would deploy after re-entry.

Martin contractors with Apollo report and reentry models (above)

1961 Langley model of what appears to be the R-3 reentry vehicle

[Martin Models 410 & L-20 1961]

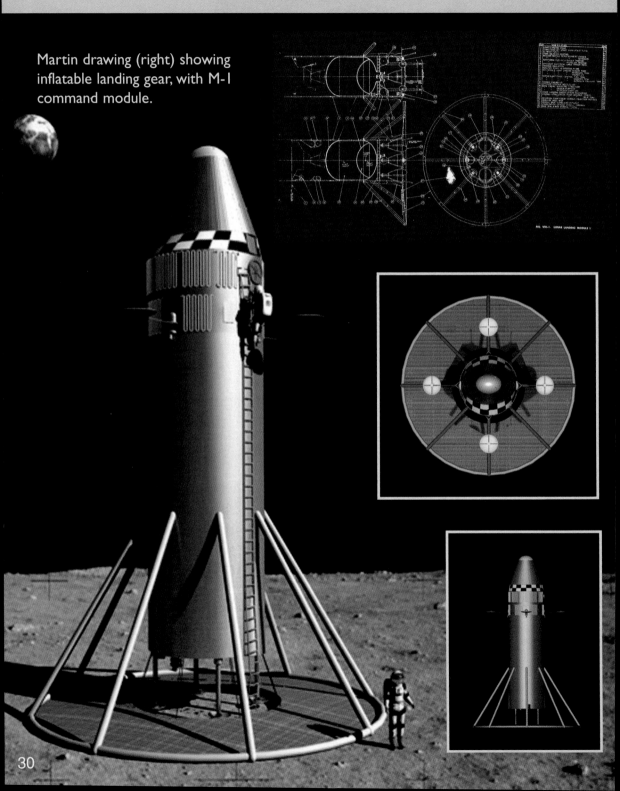

Martin drawing (right) showing inflatable landing gear, with M-1 command module.

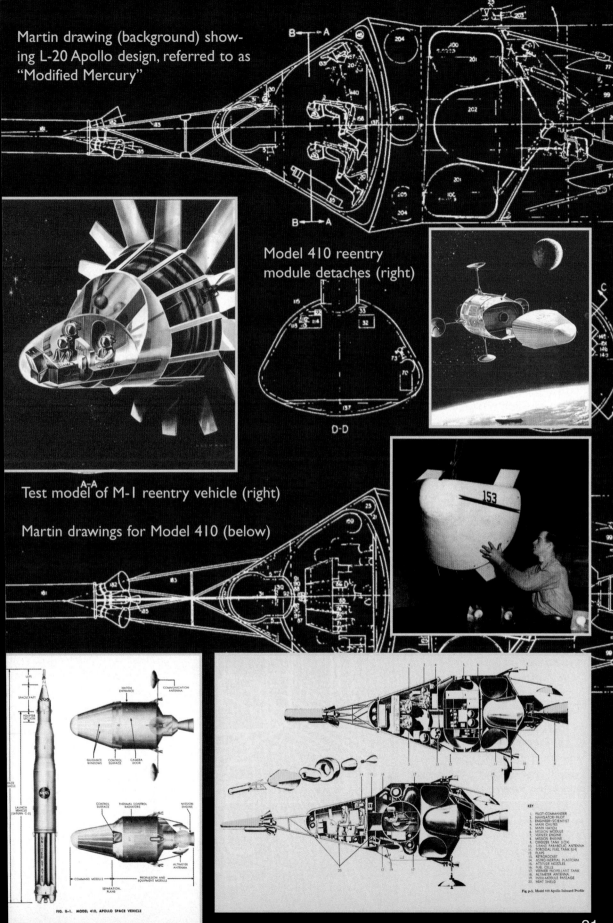

Martin drawing (background) showing L-20 Apollo design, referred to as "Modified Mercury"

Model 410 reentry module detaches (right)

Test model of M-1 reentry vehicle (right)

Martin drawings for Model 410 (below)

[USAF Lunex 1961]

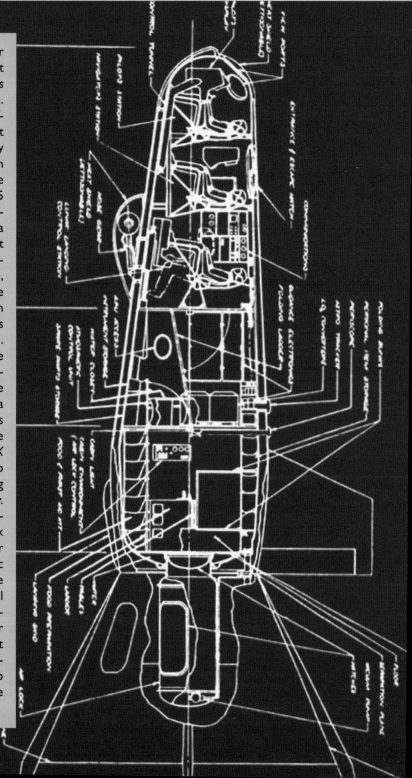

Another group agonizing over the choice between direct flight and EOR were the *Space Systems Division* of the US Air Force. *Project LUNEX* had been an offshoot of the 1950's research that ended up spawning the Mercury program. The Air Force had been conducting extensive tests in the Mojave desert with their X-15 program, and the subsequent offshoot research was generating a variety of winged spacecraft designs. This process would culminate with the *X-20* program, which would consume all of the Air Force's efforts and attention for manned space flight until its cancellation in late 1963. However, the command module for *LUNEX* would look suspiciously like many of those same shapes designed for the *X-20* (a military hypersonic weapons delivery system known more familiarly as *Dyna-Soar*). *LUNEX* was even to have been drop tested from a B-52, in keeping with ongoing Air Force policy. *LUNEX* research began immediately after the launch of Sputnik under the banner of "Lunar Observatory" and "Strategic Lunar System", but the Air Force would not present their final plan until May of 1961, immediately after the President's call for a lunar program. Had President Eisenhower not ordered the formation of NASA in 1958, Apollo might have looked a lot like *LUNEX*.

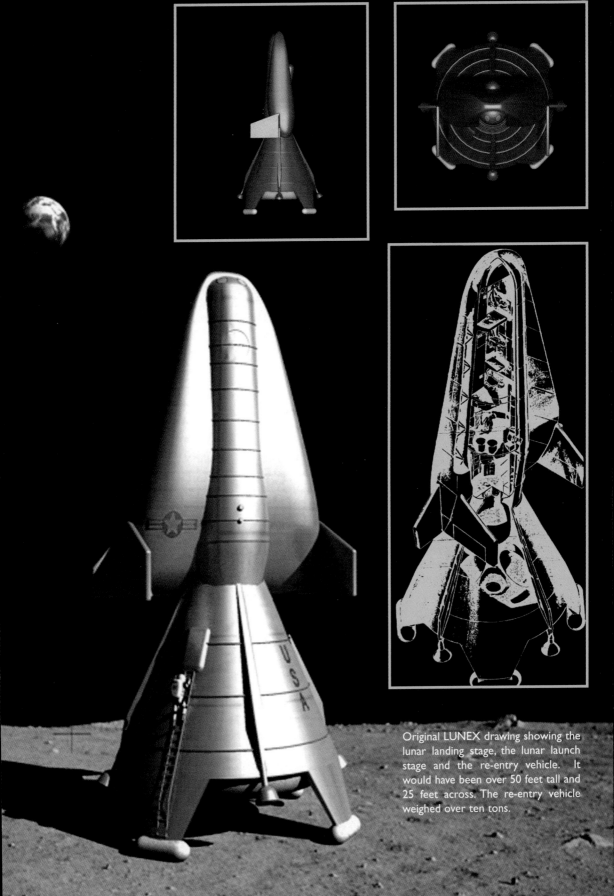

Original LUNEX drawing showing the lunar landing stage, the lunar launch stage and the re-entry vehicle. It would have been over 50 feet tall and 25 feet across. The re-entry vehicle weighed over ten tons.

[Lunar Orbit Rendezvous 1960]

In 1960 an experienced aircraft engineer at Langley Research Center, called William H. Michael Jr, presented a monograph in which he suggested that the propellant and engine required to push a lunar spacecraft back to the Earth did not need to be sent down to the lunar surface. Michael had realized that there would be a major weight saving by using this approach, but the downside was that the lunar lander would have to rendezvous and dock with the return stage in lunar orbit. This was not a new idea, in fact it began with a lecture given to the membership of the *British Interplanetary Society* on November 13th 1948. H.E. Ross the author of the paper had been with the *BIS* since at least January of 1939 when he wrote "The B.I.S. Moonship" a remarkably well conceived description of the requirements for a lunar spacecraft. In his 1948 lecture Ross explained that it made no sense to carry all of the propellant necessary for the return journey to Earth all the way down to the lunar surface. He explained that he had spent two years working on this particular problem. Although he didn't specify a modular approach, with two separate, *piloted* spacecraft (one left in orbit and one sent to the lunar surface) he quite correctly understood that the real trick to effective lunar flight was to minimize the total acceleration necessary to complete the mission. In order to land a spacecraft on the moon you need to counteract the moon's natural gravitational attraction or you will simply fall and crash. Although the Moon's gravity is only 1/6th of that here on the Earth it still has an escape velocity of 2.4 km/sec. Therefore to land safely you have to slow your spacecraft by 2.4 km/sec and then bring along enough fuel to accelerate back up to 2.4 km/sec to lift-off and escape the Moon again. By leaving the Earth-return fuel in lunar orbit, the overall requirement for lift-off mass on the Moon (and more importantly the lift-off mass here on the Earth) was reduced considerably. Here is how Ross explained it:

"With all three men aboard her, the ship, which now weighs about 65.2 tons gross, then departs from the 500 mile sub-orbit, heading for the moon. On approach to the moon, the ship is piloted into a circular orbit, say 500 miles above the satellite. Here fuel tanks weighing about 39 tons are detached and left circling in the orbit, whilst the ship descends to the moon. After touchdown on the moon, the ship will weigh about 10 tons gross. In due course the ship rises from the satellite and heads for the 500 mile orbit about the Moon. Here with the aid of radar or other search equipment, it seeks out and comes alongside the fuel tanks which were left in this orbit. This fuel is pumped into the ship, which then heads for the sub-orbit about Earth."

After some further description he concluded, "Incidentally, 'refuelling' might involve physical transfer of full tanks from one ship to another, rather than transfer of the contents alone."

Here was a tightly woven conception of *lunar orbital rendezvous*. Over a decade later William Michael would refresh everyone's memories with his own paper, and he suggested to not just leave errant fuel tanks floating in lunar orbit, but the entire piloted return vehicle, with its fuel and *engine*.

This seemingly innocuous suggestion spurred the research team at Langley to begin the push for lunar orbital rendezvous (LOR) and would be the beginning of a campaign that would continue right through into the fall of 1962. At about the same time that President Kennedy made his famous proclamation, the team at Langley, (which now included the fiercely staunch proponent of LOR, John Houbolt) presented a design for a small lunar landing vehicle that was dubbed MALLIR (Manned Lunar Landing Involving Rendezvous). The MALLIR lander was much smaller than anything conceived up to that time but it returned the lunar landing vehicle to something that looked similar to the 1940's BIS design.

[MALLIR 1962]

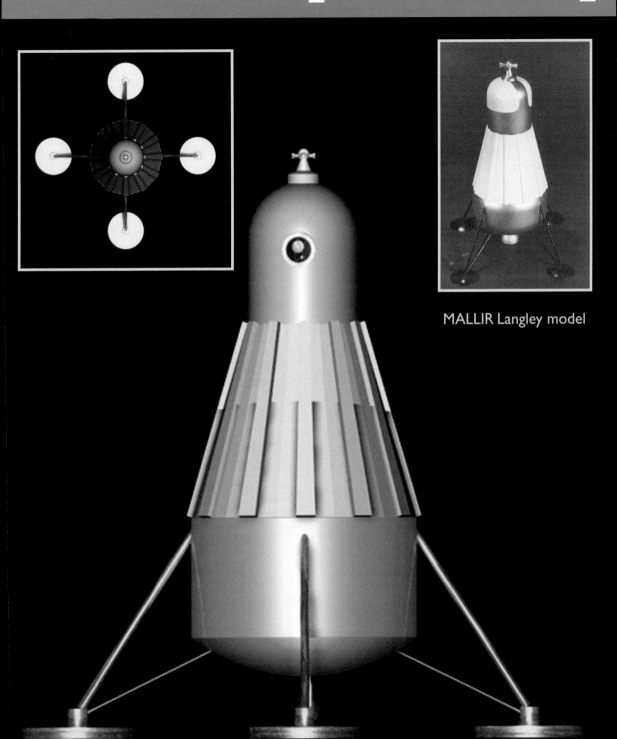

MALLIR Langley model

[Langley Landers 1961]

In August of 1961 John Houbolt gave one of many presentations about LOR to the *Space Task Group*. In attendance was Jim Chamberlin (the brilliant designer of the Canadian *Avro Arrow* fighter.) Chamberlin was heading up what would become *Project Gemini* and he had already begun to think about using the new larger *Advanced Mercury* (as *Gemini* was called at the time) to do more than just fly in low Earth orbit. Chamberlin wanted to use *Gemini* to go to the moon and he would make that exact suggestion later that year. His team's design for an accompanying lunar lander (opposite) was similar to one proposed earlier in the year by the Langley staff. It was basically nothing more than a platform placed on top of a rocket engine, on which an intrepid astronaut would stand, surrounded by fuel tanks (this page).

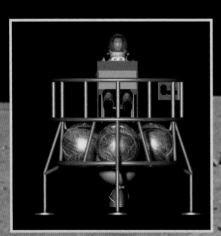

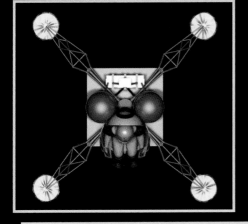
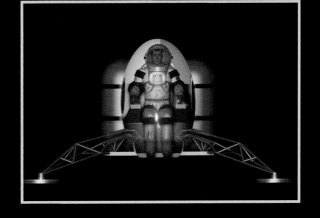

The Chamberlin team's design was sketched out that November by Harry C. Shoaf and is rendered here in color for the first time. It would likely have been launched by a Titan II in tandem with the Advanced (or Mark II) Mercury (also on a Titan II). The audacity of these shoestring vehicles is simply stunning today, but undoubtedly had they gone forward NASA would have found an intrepid pilot willing to fly them.

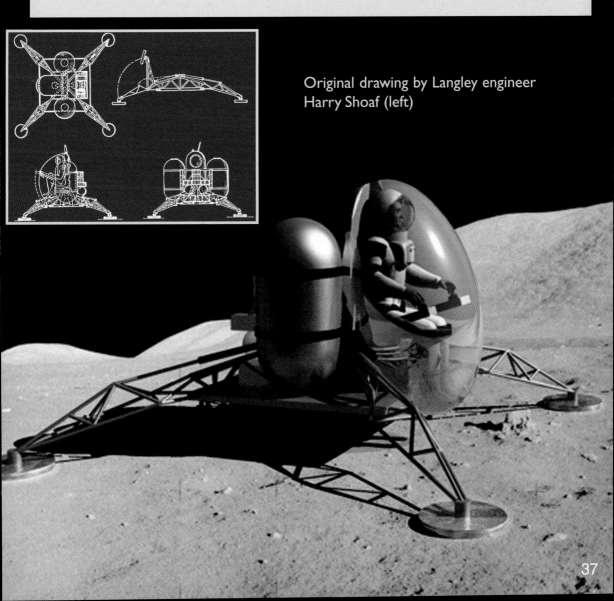

Original drawing by Langley engineer Harry Shoaf (left)

[Modules 1961-62]

The *Saturn* of 1960 was considerably smaller than that which ultimately launched men to the moon, but it was thanks to an adaptation of that old *BIS* lunar orbital rendezvous plan, that it would only require one of von Braun's giant boosters to accomplish a lunar landing. *Apollo* would now leave a piece of the spacecraft in lunar orbit while an independent modular lander would take the crew to the lunar surface. They would then return to lunar orbit where they would dock with the orbiting mother-craft. Through the Spring of 1961 the engineers at Langley had worked out that this would require ten launches of the proposed Saturn C-1. This would have taken the lunar spacecraft up in five separate parts followed by five tanker launches to take the fuel. Using the more powerful C-2 they calculated that this same process might be accomplished with only two launches if a lunar orbital rendezvous were chosen. Finally, a single launch of a Saturn C-3 was considered, but this final proposal was considered to be more than a little optimistic.

Now that Apollo would no longer be launched by a large, direct booster, but by a smaller indirect system, the lander also had to be completely redesigned. Up until this point the engineers at NASA's *Space Task Group* had been completely consumed with the basic problem of building a spacecraft that could simply take three men into space for two weeks. Little thought had been given to an actual landing craft.

Since NASA's space flight experience was minimal (the one-man Mercury capsule had barely flown at this point), the problems of designing the Apollo reentry vehicle were paramount. A spacecraft returning from the moon would be travelling at speeds in excess of 11 kilometers a second and would slam into the Earth's atmosphere like a bullet. Entirely new materials had to be devised to allow the vehicle to survive the onslaught of reentry. Because *lunar direct* was the only method that had been given any serious consideration, most of the early designs for a lunar spacecraft had concentrated on a shape that could survive the massive stresses of re-entry and then adapting that shape for a lunar landing. The lessons being learned from the early *Mercury* flights, gradually pushed the team away from the exotic flying saucers and lifting bodies. A conical shape was working well for *Mercury*, and so it was considered the best choice for *Apollo*. With so many of the NASA team steering the design toward a conical shaped spacecraft, the contractors realized it only made sense to give them what they wanted. By July of 1961 the conical re-entry cabin for *Apollo* was sanctioned and the romantic flying saucers and gliders were finally dispatched into the history books.

By October 11th of 1961 five contractor teams were ready to present their proposals for Apollo, they gathered in a hotel room in Houston and made their pitch to a NASA technical assessment panel. This panel then worked for two weeks to sift through the presentations before submitting it to the Evaluation Board. The board included Walter Williams, Associate director of Manned Spacecraft, Mercury designer Max Faget, Gemini designer Jim Chamberlin as well as three representatives from HQ and one from Marshall in Huntsville.

In November of 1961 Milt Rosen also formed his own committee to choose a method for getting to the moon. Rosen's heart was still with his giant *Nova* booster but, given only the *Saturn* to work with, he suggested adding a fifth F-1 engine to the first stage, which up until that time had only been designed for four. Von Braun had also wanted to add a fifth engine. This extra engine would give enough supplementary thrust to *Saturn* to accommodate either EOR or LOR.

NASA announced the winner of the coveted Apollo spacecraft contract in December of 1961. It had been decided that *North American Aviation* would build the "command center" and a "second component which would house fuel, electrical power, supplies and propulsion units needed for a lunar take off." It went on to say that a separate contract for a third Apollo spacecraft unit, the lunar landing sys-

Early NASA painting of an Apollo circumlunar spacecraft

tem, was expected to be awarded within three months. At this time, this proposed third unit was being referred to as a "lunar landing module". This "LLM" was to be attached to the bottom of the other two main parts of Apollo and would very likely have legs and a powerful enough engine to lower the entire three module spacecraft down to the lunar surface.

Meanwhile, this direct descent approach was bogged down with weight issues and, perhaps just as significantly, the inherent problems of trying to pilot a vertically aligned spacecraft equipped with windows that pointed upwards. This debate over direct ascent, LOR, and EOR finally came to a head on June 7th 1962. Houbolt and his team had managed to persuade the Houston center of the virtues of LOR but the political problems weighed heavily on this method. If LOR were to be adopted it would take a lot of the promised work and dollars away from some of the other NASA centers, particularly Huntsville which had a vested interest in building multiple *Saturn* rockets to support EOR, or if not that, then the gigantic *NOVA* to support *lunar direct*. Fully aware of the implications of the potential for loss of work and business for his team, and over some loud protestations from *Mercury* designer Max Faget, von Braun finally jumped on board with LOR. With the teams in both Houston and Huntsville committed to LOR, work could now proceed to produce a new lunar landing vehicle that would not need to fly anywhere but in space. This would simplify matters for everyone and undoubtedly caused a huge burden to be lifted from the shoulders of the designers at *North American Aviation* who up until that moment had been trying to build a command ship that could do both EOR and LOR.

The stage was now set for the birth of the *Lunar Excursion Module* (LEM) and the coincident death of the superbooster called *Nova*. If Houbolt's team of engineers was correct, LOR could be accomplished with one *Saturn* booster. The lander would now leave the command vehicle in lunar orbit and would go on an excursion, thus the adoption of the LEM acronym. Once the whole LOR idea was firmly entrenched the LEM got shortened simply to LM for "lunar module".

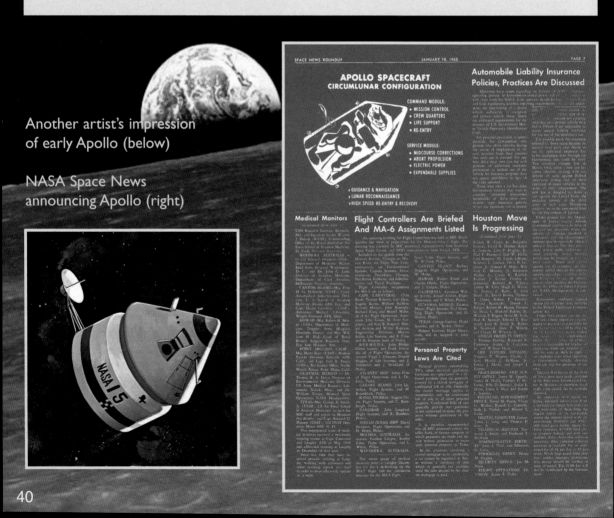

Another artist's impression of early Apollo (below)

NASA Space News announcing Apollo (right)

40

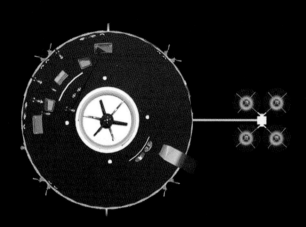

Apollo CSM side and top view

Extremely early full scale mock-up of Apollo cut away to show the interior (below and below right)

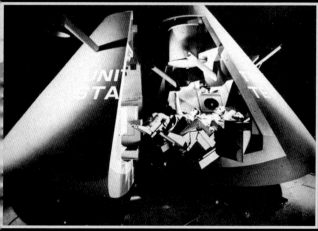
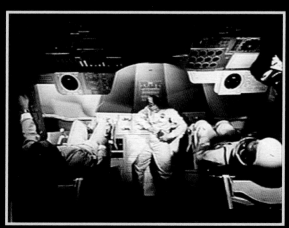

[NASA LEM 1962]

On April 16th 1962 about sixty of NASA's top people gathered at the Huntsville NASA facility to discuss the upcoming lunar program. Members of the engineering staff from Houston had been flown in, and for many of them it was their first encounter with von Braun's team. One of those in attendance was Owen Maynard, another veteran of Canada's *Avro Arrow* program. Maynard had been quietly working away since the previous summer to try and design a spacecraft that could land on the moon. The biggest obstacle he encountered was how to lay out a spacecraft in which the crew would be launched from the Earth lying on their backs, but would be able to land on the moon while looking down. It seemed an intractable problem until he remembered something he had heard while listening intently to the Langley people, specifically Jim Chamberlin. His *Avro* compatriot had favored a modular approach to solving complex systems in the *CF-105 Arrow*. This allowed repairs to be carried out without disturbing other systems. Taking this approach to its logical conclusion suggested an entirely separate module for landing.

By the time he was invited to Huntsville, Maynard had drawn up two designs for a lunar lander that would be an entirely independent module from the main spacecraft. It was obviously an offshoot of Chamberlin's mini-lander idea from the previous year, but larger and more capable. Maynard produced his slides and explained his designs to von Braun and the other NASA management. The images he showed were of a four legged spider-like vehicle. This permutation of, what would become known as the Lunar Module, was basically a conical shaped Apollo style

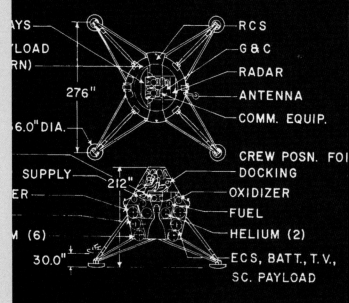

LUNAR EXCURSION MODULE
OARD PROFILE AND GEOMETRY

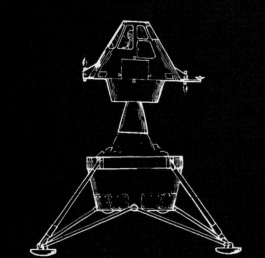

LUNAR EXCURSION MODULE
STAGING AT LIFT-OFF OR ABORT

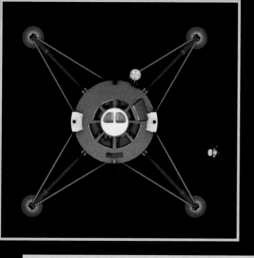
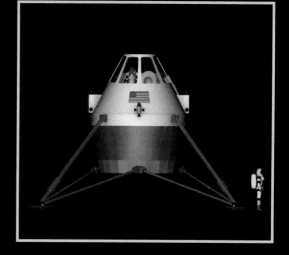

command module atop a cylindrical descent stage. It was a logical extension of the work done on Lunar Direct. It would have large windows in the front and two windows in the apex of the cone (these apex windows seem to have been hastily replaced with a hatch). At least two different arrangements of the reaction control thrusters seem to have been considered for this early design.

A variant of this particular lander was also considered for training on Earth (inset below). In place of the top windows the pilot was exposed to the sky and would have been seated in an ejector seat. Other alterations included modified landing gear, removal of the environmental controls and scientific payloads as well as the helium tanks. Finally, the main engine would have been modified for atmospheric use.

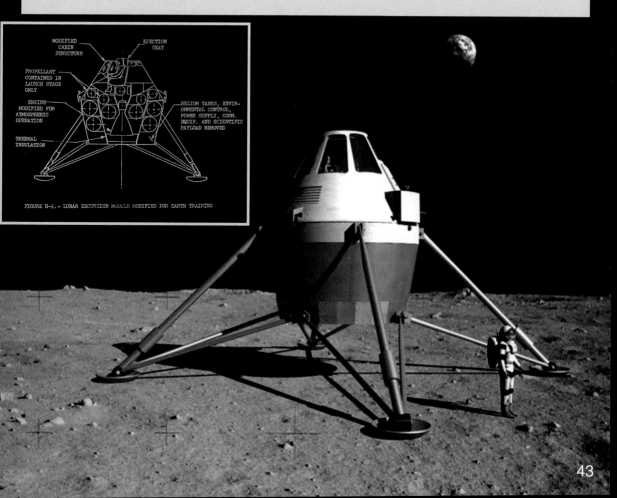

[NASA (FD) LEM 1962]

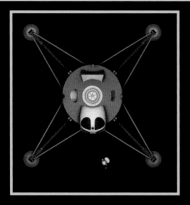
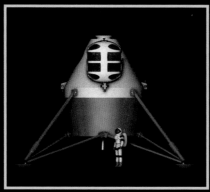
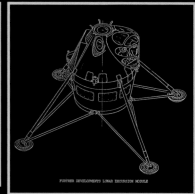

Maynard also told his audience that he had realized that the view from the windows in the side of a cone provided less than ideal lines of sight for the pilot, and so a large protrusion was added to the front of the cone in his second design. A period artist's impression of the Further Developments LEM (right) shows it with two windows still on top, this may have been an error since the hatch was already in place by May 1962. The renderings here show the proposed dual top and side hatches.

Space News NASA ROUNDUP

VOL. 1, NO. 20 — MANNED SPACECRAFT CENTER, HOUSTON, TEXAS — JULY 25, 1962

Schirra Announces Spacecraft Changes Via Telstar

Emphasis To Be Placed On Lunar Orbit Method

NASA has announced that lunar orbit rendezvous (LOR) will be the prime mission mode for the Apollo manned lunar exploration program, using the advanced Saturn C-5 as the booster. Saturn C-5 can launch 45 tons to escape velocity.

The next phase of NASA planning, research, development and procurement will concentrate on LOR, one of three possible methods of putting astronauts on the moon. The other two are the direct earth-to-moon-surface flight, and the earth orbit rendezvous. Studies of the earth-orbit rendezvous method, using the advanced Saturn with a two-man spacecraft, will be continued, as will studies of a direct flight to the moon, using such a spacecraft and the advanced Saturn. But prime emphasis will now be placed on lunar orbit rendezvous.

Studies of the Nova launch vehicle, with a weight lifting capability of at least two to three times that of Saturn C-5, will be continued but development has been deferred at least two years. Such a booster would be used for possible missions beyond Apollo.

"We are putting major emphasis on lunar orbit rendezvous because a year of intensive study indicates that it is most desirable from the standpoints of time, cost and mission accomplishment," NASA administrator James E. Webb said. "However, we have also decided to retain the degree of flexibility vital to a research and development program of this magnitude. Many of the modules and booster stages are interchangeable between the various modes open to us. If what we learn in the future dictates a further change in direction, we will be in a position to make it."

In connection with the decision to concentrate on LOR, NASA is requesting industrial proposals for the development of a lunar excursion vehicle, nicknamed a "bug," which will be carried aboard the Saturn booster with the Apollo mother ship as it is launched into orbit around the moon. The "bug" will be capable of landing two men on the lunar surface and returning them to the mother ship while a third crewman remains on board the Apollo spacecraft in lunar orbit.

Plans call for the use of a two-stage Saturn (configuration B) using the present eight-engine Saturn first stage, and the high energy S-IVB stage already under development, for early flight tests in earth orbit in the mid-1960's. These flight tests will be utilized to perfect maneuvers in earth orbit with minimal fuel loads. Saturn C-1B will develop sufficient thrust to put 16 tons into earth orbit. Saturn C-5 will put 120 tons into earth orbit.

An in-depth study of an unmanned lunar logistic vehicle to support the lunar exploration program will be begun immediately.

Members of NASA's Manned Space Flight Management Council, under the chairmanship of Manned Space Flight Director D. Brainerd Holmes, recommended LOR unanimously for four reasons. It provides a higher probability of mission success with equal safety some months earlier than other modes, and within the national goal period of this decade. It will cost 10 to 15 per cent less than other modes, and requires the least amount of technical development beyond existing commitments.

The Council is composed of the directors of the Office of Manned Space Flight in Washington, D. C. headed by Holmes; MSC here in Houston headed by Dr. R. R. Gilruth; Launch Operations Center at Cape Canaveral headed by Dr. Kurt Debus; and Marshall Space Flight Center in Huntsville, Ala. under Dr. Wernher Von Braun.

As presently envisioned, LOR would require a single launch of a Saturn C-5 boosting a 13-foot diameter, three module spacecraft. The spacecraft would include a five-ton, 12-foot tall command module housing the three crewmen; a 23-ton, 23-foot tall service module providing mid-course correction and return-to-earth propulsion; and a 15-ton, 20-foot tall lunar excursion vehicle. The three modules would be placed in lunar orbit as a unit. Two astronauts would then transfer to the lunar excursion vehicle and descend to the moon while the Apollo

(Continued on Page 2)

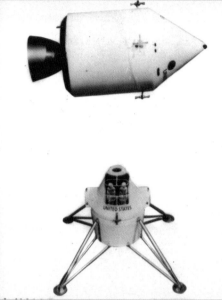

PRESENT CONCEPTION of landing on the moon using the lunar orbit rendezvous method, now designated the prime mode for Project Apollo, will include an excursion vehicle (shown here in small model form) with two astronauts aboard descending to the surface as the command and service modules remain above in lunar orbit. (See additional pictures on page 3.)

Both pictures on this page show a model of LEM that appears to be half way between the two Maynard LM designs of 1962. They show a distinctive cockpit bulging forward, like the FDLEM but still retain the large RCS pods on each flank. The cockpit also seems to be more rectangular than the one seen opposite.

The space age turned one of its more spectacular sides to view Monday afternoon as Astronaut Wally Schirra, soon to orbit the earth, spoke to the better part of its people via a package of electronics which was already doing so.

Part of the 45-minute show on Americana which was broadcast through American Telephone and Telegraph's experimental communications satellite, Telstar, was devoted to changes in the Mercury spacecraft which will lift Schirra into orbit late in September.

The program was beamed to most countries of the world, with the exception of Japan.

One day soon, via Telstar perhaps, the nations of Europe will join us for a live coverage of a man in orbit.

Schirra was speaking from Hangar S at Cape Canaveral, where he is hard at work on preparations for the MA-8 mission.

"My flight plan calls for up to six orbits . . . nine hours in space," Schirra explained. "There are only a few equipment changes.

[text obscured by overlay]

When asked if the decision meant that Slayton would never fly in a spacecraft, Purser replied "not necessarily."

His ailment is known as atrial fibrillation, a periodic lack of rhythm in the heart

Astronaut Donald K. Slayton

pered by a shortage of fuel for the attitude control rockets which at one point was considered serious.

Schirra will take with him

(Continued on Page 2)

45

[NASA LEM 1962]

At the same April meeting in Huntsville, von Braun was seen playing with a large model for a lunar lander which he showed to visiting NASA staff and various contractor representatives. This alternate three-legged NASA design, also for an entirely independent landing craft, would blossom into a full size mockup within five months. Little seems to be known of this alternate design but it seems that it also originated at the Houston center. A rare (low rez) color photo of the mock-up can be seen at lower right, a better shot on page 53.

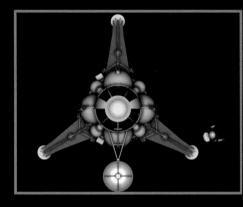

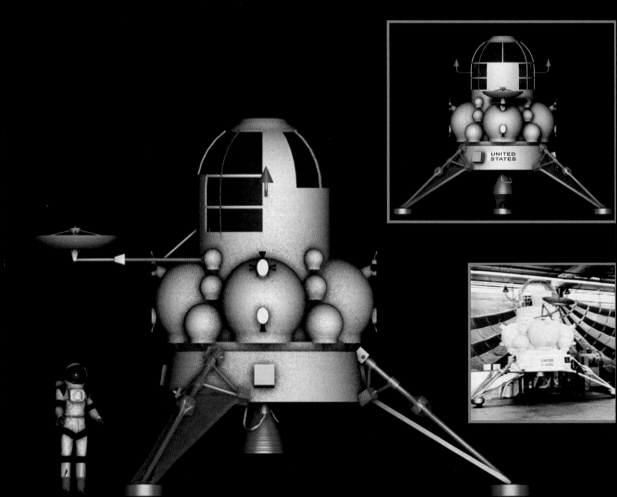

Wernher von Braun in April 1962 examines an early NASA lunar lander with Joe Shea, Ernst Stuhlinger and John Glenn.

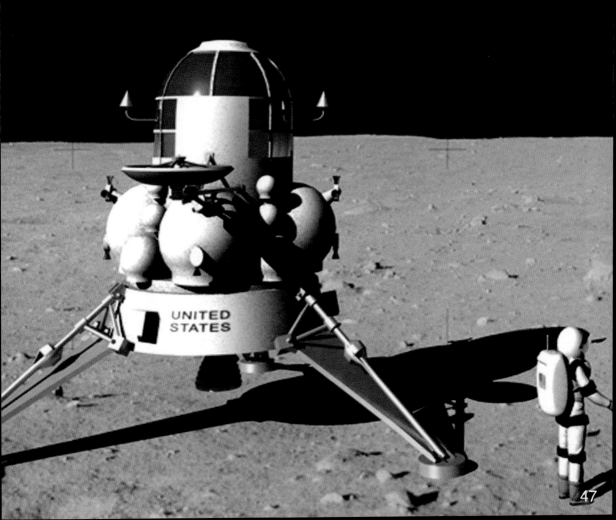

[MISDAS and CM variants]

By this time *North American Aviation* had been thrashing away to complete the Apollo command/service module aggregate which would remain in lunar orbit, and it had already evolved into something approaching its final familiar configuration. The CSM had gone through many tests, one of which was to try and determine the best way for it to land back on the Earth. As late as 1967, the MISDAS study was still considering at least eight different landing architectures for the Command Module. Another logistics vehicle under consideration would have been made by modifying the Apollo command module. The CM logistics vehicle would have been connected to a small propulsion module which in turn would be connected to a consumable cargo module and a special cargo module. In many respects this CM logistics module looked very much like the earlier Lunar Direct return module. Studies continued on this Apollo off-shoot until at least the summer of 1967. Below are the four final designs from 1961 that were rejected in favor of NAA's design.

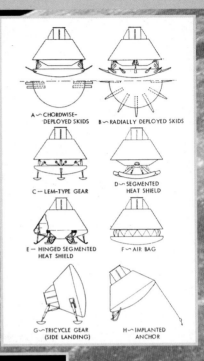

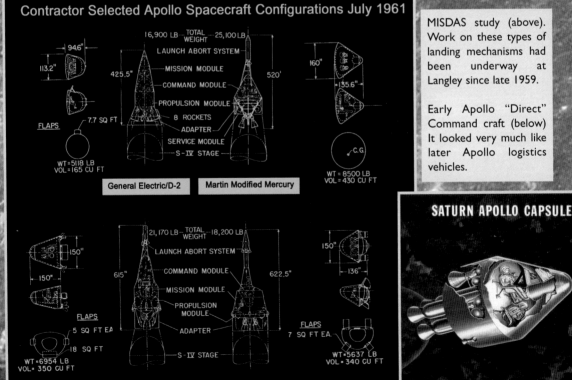

MISDAS study (above). Work on these types of landing mechanisms had been underway at Langley since late 1959.

Early Apollo "Direct" Command craft (below) It looked very much like later Apollo logistics vehicles.

[Lunar Orbit Rendezvous 1962]

On July 11th 1962 NASA's Joe Shea made a presentation of the proposed LOR architecture to the media, using an array of large models that included Maynard's bulging, glass-cockpit, conical LEM, and the now familiar Apollo CSM. He was able to make comparisons between the LOR models and the much larger and soon-to-be-extinct *Apollo Direct* models. Here is what he said:

> SHEA: I think it probably appropriate that we concentrate primarily on the LOR mode. I think these models will give you a feel for the upper stages.
> The launch vehicle, as has already been mentioned, is the C-5 (Saturn V) consisting of the five F-1 engines S-IC first stage. Five J-2 engines S-II second stage, and the S-IVB escape stage. The basic mission mode calls for a single launch of the vehicle from the pad at the Cape. It will require the burn of the S-IC stage, the burn of the S-II stage and a partial burn on the S-IVB stage or third stage to put into earth orbit the S-IVB, the lunar excursion vehicle, the service module and the command module.
> This is the payload then that will exist in earth orbit. It will have to go around at least a half revolution in order to get to the proper launch window point, check out the spacecraft, and see that we are ready to actually commit to the mission.
> At the time that we commit to the mission the S-IVB will burn again to provide the additional velocity increment to inject the spacecraft on the trans-lunar trajectory. Once we are on the trans-lunar trajectory the total spacecraft weight will be in the order of 85000 pounds or thereabouts. The injection capability of the launch vehicle is in the order of 90000 pounds. So at this stage in the program we have some comfortable weight margin between spacecraft requirements and launch vehicle capabilities.
> After injection we don't want to carry the S-IVB as an integral part of the system and it is necessary to use the propulsion in the service module for possible aborts. The operational mode then consists of moving the service module/command module combination off the S-IVB and lunar excursion module, opening the fairings coming back around re-orienting the command module/service module so that we actually mate the command module/service module combination with the lunar excursion module and then once that operation is accomplished the S-IVB is dropped away.
> This particular set of model stages is a little bit easy so let me uncouple them for a minute. The configuration which we then have on the way to the moon is effectively this configuration.
> We will be able to check out the lunar excursion module and determine that its subsystems are working, the actual midcourse guidance corrections we need to keep us on this trajectory will be determined by on-board guidance equipment in the command module itself. The propulsion will be provided by the service module propulsion system.
> When we get to the moon approximately 72 hours later the service module propulsion will burn. This entire configuration will drop into lunar orbit.
> In lunar orbit we will then have again this assembly. The orbit will be approximately a hundred miles above the lunar surface and will be roughly in an equatorial band some plus or minus ten degrees latitude from the lunar equator.
> After determining that all the subsystems are working and that we are ready to commit to the mission, two of the three astronauts will transfer from the command module to the lunar excursion module. Once they are transferred we will then be using the propulsion aboard the lunar excursion module to put the excursion module on a trajectory which has the same period as the circular orbit of the command module/service module combination but has a much lower perigee of approximately 50000 feet. This will enable us to go down and in effect examine, from an altitude of something like ten miles, the intended launch site.
> The equal period nature of the orbit means that it is sort of natural that these two vehicles will come together again once each orbit so you have a natural position for re-rendezvous if for any reason you want to abort the mission or decide not to commit down to the lunar surface.
> Once you decide to commit to the lunar surface you then again burn the engine on the lunar excursion module and it will then provide approximately 7000 foot per second velocity gain to bring you down to a hovering position with respect to the surface. You now have your landing legs extended. You have the capability to hover for something like a minute, to translate the vehicle something like a thousand feet to actually pick the point of touchdown and then the vehicle will land on the lunar surface.
> The trajectories again can be constructed in such a way that all during this retro-maneuver, hover and maneuver and landing maneuver on the lunar surface the command module/ service module, with the one astronaut aboard up here, will always be in line of sight and radio communications with the lunar excursion vehicle.
> All during this descent phase if again for any reason an abort is desired, there are a very simple series of trajectories which

will allow this vehicle to abort and rendezvous with the mother craft itself.

I think you can see some of the features and characteristics of the LOR just by looking at the size of the landing vehicle itself. Basically we were able in this mode to design a space vehicle specifically for operation in the vicinity of the moon to provide a reasonable amount of glass area so that the landing maneuver can be under visual control of the astronauts and that the actual touchdown site can be given a reasonable observation before we touch down.

In addition the size of the landed vehicle is such that landing gear restrictions are somewhat minimized. The entire vehicle is just optimized in effect for the landing maneuver itself.

That also lets us incidentally, optimize the command module for re-entry into the earth's atmosphere and essentially optimizing that configuration requires a minimum amount of glass, a shape which is in effect a shape like the Mercury capsule. At this point in time a positioning of the astronauts so that they are able to withstand entry Gs, so that their normal position is in effect laying on their backs as far as this configuration is concerned, rather than the vertical position that we would like to have them at for landing.

After the mission is accomplished and we have something like a two to four-day stay and exploration time on the lunar surface we decide to commit then to the return capability. We stage the lunar excursion module and leave on the surface the tanks required to carry the fuel for the landing, the landing gear itself and this landing stage in effect becomes a launch pad, a lunar Canaveral.

At an appropriate time, we have a launch window there of something like six or seven minutes. With the orbiting spacecraft coming up overhead — as a matter of fact about two or three degrees just behind you — you ignite the engine in the lunar excursion module climb up a trajectory which enables you then to rendezvous with the mother craft. All during the ascent maneuver we have radar contact and visual contact between the lunar excursion module and the command module/service module combination. We provide a capability for making the rendezvous from either the lunar excursion module or from the command module/service module combination. This in effect being a critical maneuver we are able to provide complete redundancy, in fact in some cases double redundancy, in terms of sensors and control systems aboard both the excursion module and the command module/service module combination.

Assuming everything works we come up here, make a mid-course correction about half-way up the ascent trajectory. A bit further on when the two craft are about three miles apart, using the radar data and possibly the optical data, the lunar excursion module will reorient itself, bring itself to a position where it has a very small velocity error and very small linear displacement from the mother spacecraft, and then under the control of the astronauts the two craft will again be joined.

The astronauts that have been on the lunar surface then transfer back to the command module — the lunar excursion module will then be left in lunar orbit, the service module will burn to provide the propulsion to get out of lunar orbit to put you on the return trajectory.

Coming back the guidance system on board, plus ground corrections are used to determine the corrections necessary to get you into the re-entry corridor, with propulsion again provided by the service module.

Just before re-entry you drop off the service module, the command module reorients for re-entry, and the mission is completed.

Joe Shea demonstrates the models for LOR, July 11th 1962

[Boeing LEM 1962]

Just a few weeks after Shea had made his elaborate demonstration using these large models, NASA issued its first request for proposals for a lunar landing vehicle. The date was 25th July 1962. *Boeing's* proposal included two different configurations for a descent stage, one with four legs and one with three. Perhaps they were keeping their options open by responding to the three-legged model that had appeared three months earlier in Huntsville. Regardless, the *Boeing* LEM was somewhat smaller than that of the competition. It was submitted on September 4th 1962 and the four-legged version stood just over 19 feet high, carrying over six tons of propellant. Its dry weight was just over 5 tons (for comparison purposes, the final LM dry weight was five hundred pounds more, and stood three feet taller.) The alternate version, with only three legs, had a dry-weight over 3000 pounds lighter. This version used cryogenic engines (i.e. H2 and LOX) and a single RL-10 engine for landing and take-off. The rendering shown here is of the more practical, hypergolic-propelled, four-legged version.

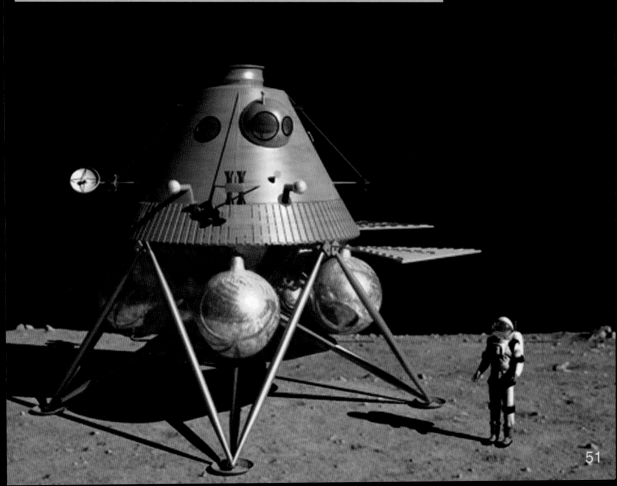

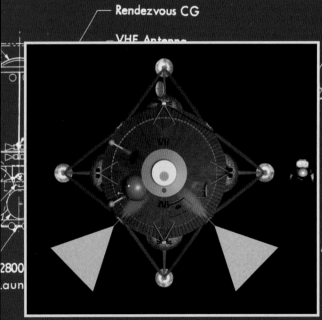
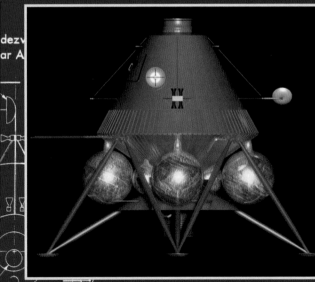
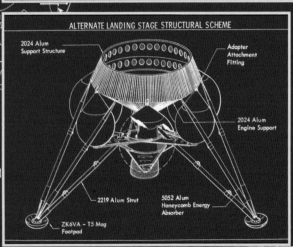
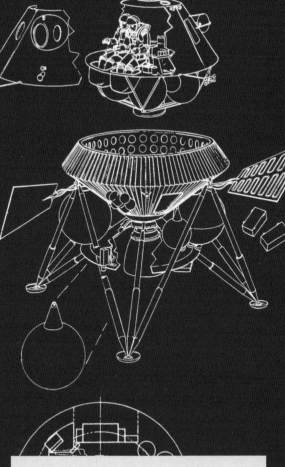
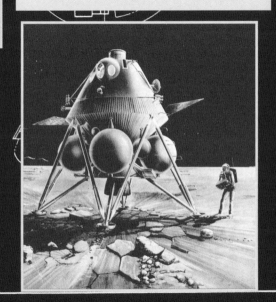

Boeing submission for an Apollo lunar lander from 1962 (above) and the three legged cryogenic version (below left) Artist's concept showing ladder on leg (below right)

LOR had now become the accepted paradigm within NASA but there were still some nay-sayers who would not let go of *Lunar Direct*. On Tuesday September 11th 1962 President Kennedy was in the middle of a swift two-day tour of the NASA centers. During the afternoon he was being guided around the Marshall Space Flight Centre when he became aware of just how deep-rooted the argument had become over LOR, EOR and *Lunar Direct*. His appointed representative, Jerome Wiesner, had sparked a mildly embarrassing argument while von Braun had been guiding the President around the Huntsville facility. Wiesner was dead set against LOR and had been fighting it for over a year. When von Braun mentioned it to the President, Wiesner interrupted him, telling the President that the idea was "no good". NASA administrator James Webb stopped Wiesner in his tracks, but not before the media became aware of the internal strife.

The following morning Kennedy made a speech to the workers at the Manned Spacecraft Center in Houston while standing in front of a full-size mock-up of a lunar landing craft, clearly designed for Lunar Orbital or Earth Orbit rendezvous. It was identical to the model that von Braun had been studying the previous April and bore almost no relation to Maynard's sleek designs. This bulbous looking creature presented a truly bizarre aspect to the visiting dignitaries. Fuel tanks were arranged around the outside perimeter, and the flight deck was surrounded with a large array of windows that allowed potential pilots to see in many directions. This unique three-legged design inevitably appeared and disappeared during the span of 1962.

Von Braun escorts President Kennedy around the Huntsville facility (left)
Kennedy makes speech at JSC in Houston (below left)

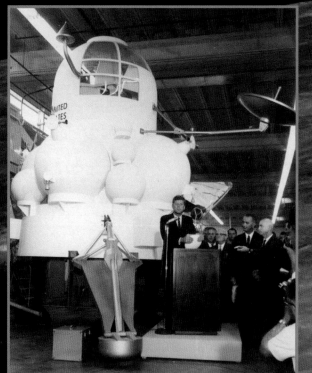

[Apollo 2-man Direct 1962]

Wiesner was still applying pressure, through the President's Scientific Advisory Committee (PSAC), to overturn the decision to use LOR. NASA's Joe Shea resignedly committed more money, yet again, to see if a smaller two man *Apollo* could make it to the moon using the *Direct* method, launched by a single *Saturn V*. This arrangement would have used a throw-away third stage on the lander. The original concept for this "crash" stage had come from *Mercury* designer Max Faget in 1961, in response to a political argument over the distribution of work-loads between the NASA centers. Faget was a superb engineer and he had found several excellent technical reasons for this extra stage, and so, for a short time it had become one of the many considered proposals. Now, in October of 1962, it would be resurrected by *McDonnell* in response to the criticisms coming from the PSAC office. On 31st October 1962 *McDonnell* presented *Lunar Gemini* and *Apollo 2- man Direct*. Both systems would include a third module, called a "retrograde stage" since it would be used to slow the vehicle as it approached the moon. It

Original drawings of the cabin from McDonnell Apollo 2-man Direct

LUNAR LANDING POSITION

QUICK OPENING HATCH

EMERG (BLOW HA

FIGURE 2-4

NAVIGATION POSITION

would have a throttleable engine with thrust from 26,500 pounds down to 2,650 pounds, would be powered by LOX and liquid hydrogen and stand over 16 feet tall. This would be mounted beneath the "terminal landing stage", which had the landing legs, and was powered by nitrous oxide and monomethyl hydrazine, using yet another throttleable engine. This stage would stand six and a half feet high and would in turn carry the "service module" which was to be used for both lunar ascent and as the main engine to return the command module to Earth. Finally, the whole rig would be capped off by either a conical modified *Apollo*, with two astronauts, or one of three slightly different versions of the *Gemini* command module. This whole system was marginal in performance and considered risky by the team in Huntsville and would be dropped in the final selection round.

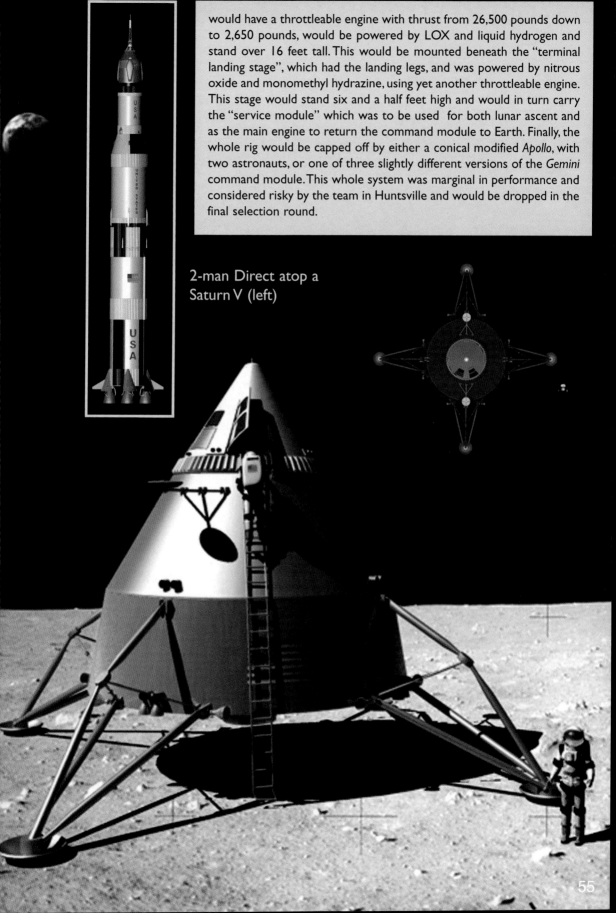

2-man Direct atop a Saturn V (left)

[Lunar Gemini 1962]

LUNAR GEMINI I INTERIOR ARRANGEMENT

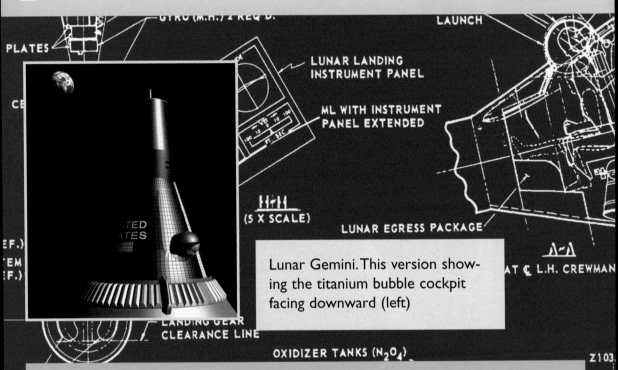

Lunar Gemini. This version showing the titanium bubble cockpit facing downward (left)

The three versions of *Lunar Gemini* included one with an almost unchanged basic command module, a second version that would use the proposed paraglider landing system on return to Earth, and finally a third version in which the ejection seats were replaced with an LES tower and a triple parachute system.

All of these systems still had the fundamental drawback of not allowing the astronaut an unobstructed view of the lunar surface during descent. Both the Apollo and Gemini command module had windows that faced in the wrong direction. Discussions ensued about using mirrors and TV cameras to solve the problem but, in the case of *2-Man Apollo*, the capsule would be redesigned to include a large hatch with two windows, one of which would be large enough to allow the pilot to see downward. In the case of the *Lunar Gemini* proposal, more creative solutions were required. Three different canopy configurations were proposed for landing, including an open hatch with a zippered canopy, a titanium reinforced bubble, or simply a rear-view mirror. The first of the three would allow for the least redesign work but would literally require the pilot to lean out of the hatch while flying down to the moon. In the rendering here you will see the version that replaced the left hatch with a titanium and glass bubble that allowed the pilot to kneel and look down towards the moon while flying the vehicle.

Both *2-Man Apollo Direct* and *Lunar Gemini* were to be launched on a *Saturn V* with an assortment of aerodynamic fairings covering the legs and other protrusions required by the design. These two exotic creations, were *still* not the last-gasp for *Lunar Direct* even though they were only considered for barely a week. On November 7th 1962 the contract to build the lunar landing craft was awarded to *Grumman* who had spent almost two years unofficially working on design ideas.

Lunar Gemini atop a Saturn V with an escape tower (right). Note fairings to cover landing legs.

LEM LUNAR LANDING STAGE

[Grumman LEM 1962]

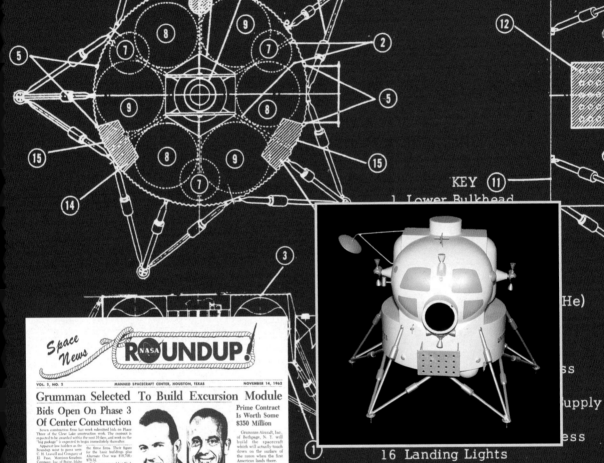

KEY
1. Lower Bulkhead

16 Landing Lights
17 TV Camera

Just over two weeks after winning the contract *Grumman* revealed their first design for a lunar landing craft. Strangely they had increased the landing gear from four to five, short, rigid legs. The cabin was basically a spherical shape and the crew would egress using a front hatch stepping out onto a hydraulic platform. The first *Grumman* LEM was presented to the world on November 23rd 1962. This preliminary five-legged design would remain (at least publicly) the preferred shape until the following summer.

58

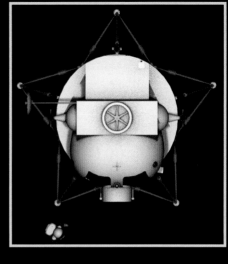
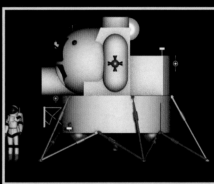

LEM LUNAR TAKEOFF STAGE
ORIGINAL CONFIGURATION & DATA SUMMARY

Apollo astronaut Jim McDivitt examines a model of the Grumman ascent stage

Space News article (below) showing Grumman facility for building the LEM (below)

Grumman brochure for LOR (right)

Grumman Aircraft Engineering Corp. Builds LEM Which Will One Day Touch Down On The Moon's Surface

Page 48

[STL Direct 1963]

Lunar Direct was still rearing its head in the form of a late proposal from *Space Technology Laboratories Inc* of Redondo Beach California. Joe Shea had been STL's Space Program Director before assuming his post at NASA. Although the LOR architecture was clearly set in stone by this time; *STL* gave one last shot at designing a 123 inch diameter 2-man Apollo capsule that used similar features to the *McDonnell* version that had appeared a few months earlier. What seems to have been unique to the *STL Apollo* was the total absence of landing legs, instead this version would land on a series of short pads. This proposal also discussed a larger 154 inch, 3-man version that might be able to function for 14 days and had a dry weight for the command module that was over a ton lighter than the *North American* Apollo proposal. The entire system would have operated on cryogenic propellant and would have still been launched on a *Saturn V*. This almost certainly was conceived at the same time as the TRW/STL LLV series (see page 145). As interesting as this late proposal appears, it could not derail the LOR plans at this late stage. Just eight days after the STL proposal was submitted, on February 9th 1963, NASA chief Robert Gilruth presented the five-legged *Grumman* LEM to congressmen and dignitaries. In the eyes of NASA, *Lunar Direct* was finished.

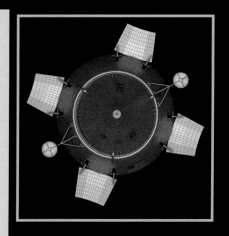

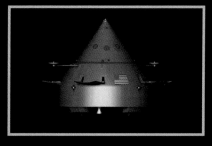

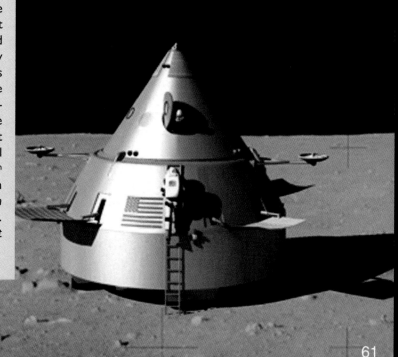

[Grumman LEM 1963]

By May of 1963 the basic five-legged shape was still in place but was besieged with problems due to the weight of the windows. So much glass presented a predicament, and just shrinking the windows wouldn't solve the dilemma. The spherical shape was inherently inefficient and unnecessary, and by this time the entire weight of *Apollo* was in danger of surpassing the lifting capability of the *Saturn V*.

In July of 1963 the spherical cabin was ditched in favor of a more practical cylindrical shape, but this was still not enough to fix the weight problems.

Four windows were still too heavy, and they also allowed too much heat to enter the cabin. A cylindrical cabin resolved some of these problems and triggered an ongoing process of shaving bits and pieces from the overall shape of the vehicle, but two of the windows still had to go. This version of the LM appeared in drawings with the cylinder partly sheared away below the two windows, and with and without panel walls on the descent stage.

At the same time it was decided that five legs was an awkward and inefficient design, and so the overhaul continued. Rosen and Schwenk had been pushing four legs for the lander in their original 1959 *Lunar Direct* plan. Most of the proposals in the interim had continued with that choice until *Grumman* were issued the contract in November of 1962. The only notable exception seems to be the three-legged version that popped up at the meeting in April of 1962.

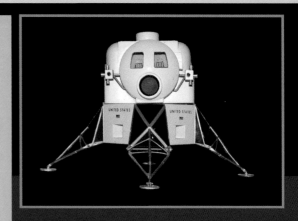

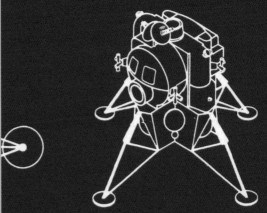

Grumman drawings for LEM circa 1963 showing exposed descent stage tanks Grumman model (above) showing tanks covered.

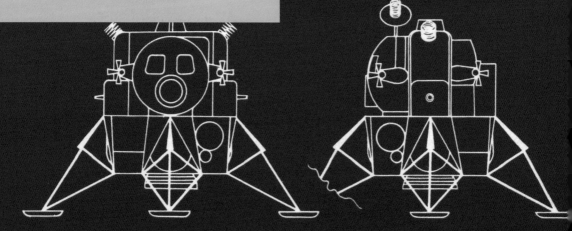

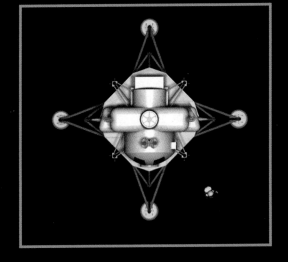
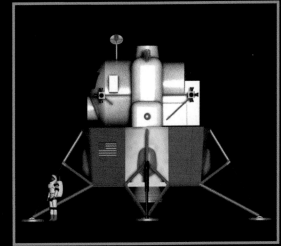

Rare Grumman LEM familiarization manual (left) and Bonestell interim painting of 1963 LEM still with the extra windows (right) Courtesy Bonestell Space Art

[Bendix LEM 1963]

Four legs restored the symmetry that Maynard had adopted for the modular lander at NASA a year earlier, but the overall vehicle weight had increased considerably since then. This meant that the diameter of the footpads needed to be increased to take the load, but the stronger legs would now no longer fit inside the *Saturn* shroud, they therefore needed to be completely redesigned. At this point an assortment of folding mechanisms were introduced. The *Bendix* corporation were one of the companies to send through a proposal that July. This created a momentary hybrid of the *Grumman* spherical fuselage with *Bendix*'s extremely large and exotic landing gear. At least one of the problems with the *Bendix* design would seem to be the sheer length of the legs. Their design would place the egress hatch almost 20 feet off the ground, which would have made *Grumman*'s hydraulic egress platform redundant. This "wide-track" landing gear was designed to accommodate slopes of up to 15°. In contrast to regular gear, designed for 7°, this placed the centre of gravity below the line of action of the lower struts. At this point not one of the LEMs, either pre-*Grumman* or otherwise, had shown a position for a ladder. But at this point the leg design was not the only thing that had not been finalized.

Bendix contractor model showing wide landing gear. Legs folded and the honeycomb footpad (below right)

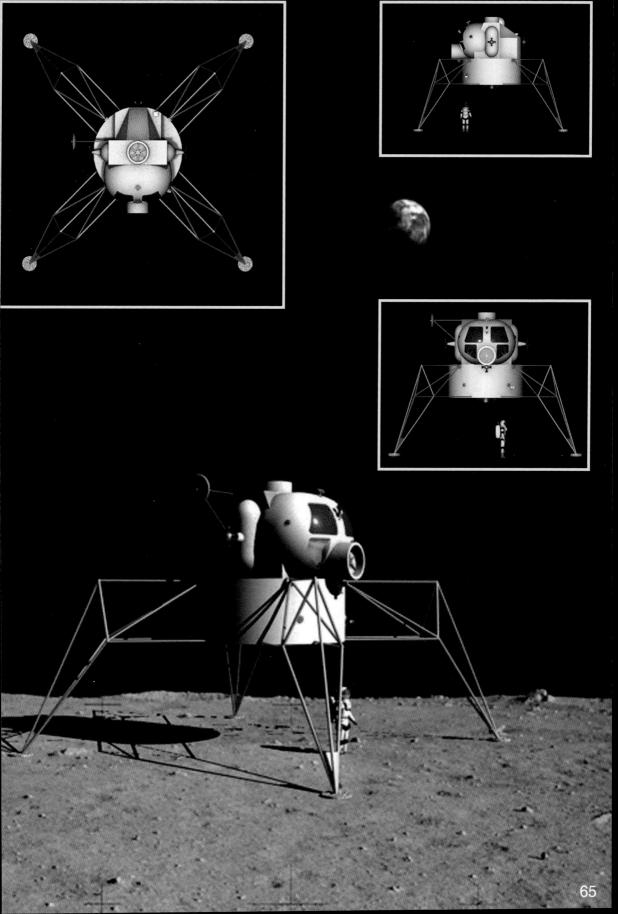

[Grumman LEM 1963]

Within three months of this major redesign, the *Grumman* team were forced into yet another drastic rethink. The ascent stage had, up until that point, two oxidizer tanks and two fuel tanks. This allowed for a symmetry in the design, but the dual tank system added weight and complexity, and so it was decided to reduce the tanks to one of each. This in itself introduced another problem, the fuel and oxidizer had different densities so it would no longer be possible to situate them tidily on either side of the vehicles' center line. Symmetry was now abandoned in favor of practicality, the LEM would, from October 1963 forward, appear to be off-centered. At about the same time it was realized that the crew didn't need seats inside the LEM. If the seats were removed the pilots could get closer to the windows and have a better field of view. However, the rounded windows did not offer the optimum viewing angle so they were replaced with triangular windows that were angled downward. The LEM was now approaching its final recognizable shape.

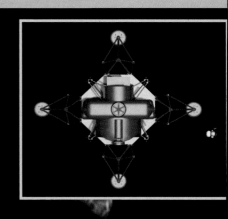

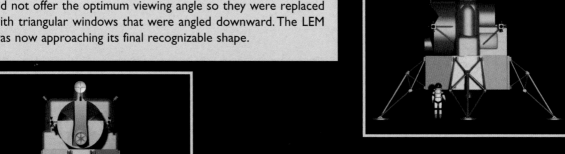

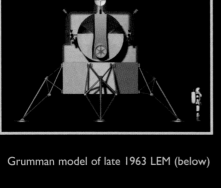

Grumman model of late 1963 LEM (below)

[Lockheed LTV/LBV 1963]

As the LM was gradually taking shape, the contractors began to think about lunar surface exploration. One of the first systems proposed was the Lunar Ballistic Vehicle which came from Lockheed's Extended Lunar Operations study. The design for the LBV came from Lockheed engineers led by George Honzik. This 12 foot diameter two-man pod was to be used for up to 200 mile-range exploration hops. Powered by hypergolics this vehicle was designed to supplement Lockheed's rather complex Lunar Traverse Vehicle. Both vehicles would have been used to support a large long term lunar base delivered to the surface by a Lunar Logistics Vehicle similar to those on page 145.

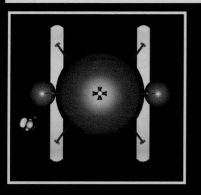
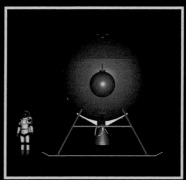
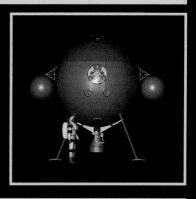

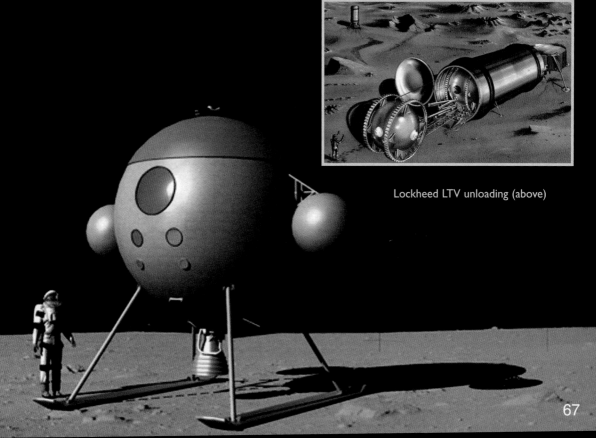

Lockheed LTV unloading (above)

67

[Langley LLRF 1963]

While *Grumman's* team was struggling to shed weight and finalize their design, the engineers at Langley were trying to figure out how the pilots might train to fly the final landing vehicle. By November 27th 1963 Langley's experimental LEM simulator was nearing completion. It looked like an old *Bell* helicopter cockpit placed on top of a mass of pipes and tubes, supported by four legs. This ungainly contraption was to be dangled from a huge scaffolding and would, in an ideal situation, provide a reasonable facsimile of what it would be like to fly the LEM. Once completed the *Lunar Landing Research Facility* (LLRF) at Langley could allow pilots to fly up to 360 feet downrange, 500 feet cross range and 180 feet vertically, all while suspended from a servo-controlled hoist system. Five-sixths of the weight of the vehicle was counteracted by the two vertical cables. This apparatus was the main training system available to the astronauts until the arrival of *Bell* Aircraft's *Lunar Landing Research Vehicle* (LLRV) at Dryden Research Center in California, in April of 1964.

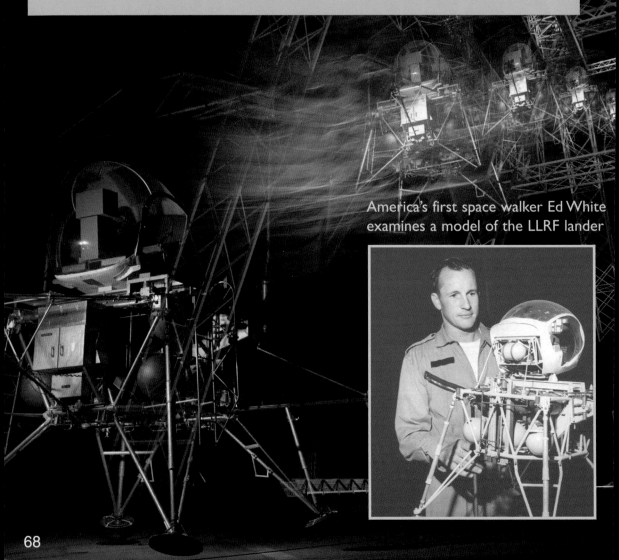

America's first space walker Ed White examines a model of the LLRF lander

The LLRF lander is prepared for flight

Langley engineer examines a model (bottom left)

Space News announcing the LLRF (left)

Early 1961 proposal for LLRF with Apollo Direct (bottom right)

[Bell LLRV/LLTV 1963-64]

The LLRV (Lunar Landing Research Vehicle) was originally designed in 1963 by John Allen Jr. and Kenneth Levin. The patent filed shows a markedly different design to that which flew a year later. This original version had the pilot's seat mounted directly over the main engine and the entire structure gimbaled around that centre engine. (right) This version seems to have been built full-size at Langley.

The LLRV made its first flight at the hands of test pilot Joe Walker on 30th October 1964. These early vertical take-off trainers would not evolve sufficiently for use by prospective Apollo astronauts until the end of 1967. There would be two LLRVs and then by mid-1966 *Bell* was given a $7.5 million contract to produce three of the final *Lunar Landing Training Vehicles* (LLTV). Three of the five vehicles were destroyed in crashes. The remaining LLRV (#2) is on display at Dryden Research Centre in California, while the one surviving LLTV (#952) is hanging in the lobby of Building 2 at the Johnson Space Center in Houston.

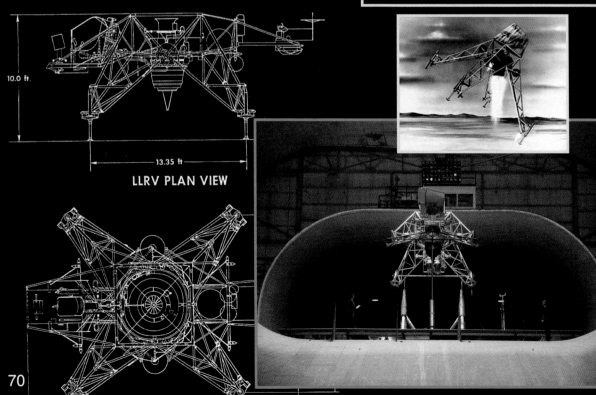

LLRV PROFILE VIEW

LLRV PLAN VIEW

The LLTV (Lunar Landing Training Vehicle) (right) had different instruments and controls to more closely approximate the LM. While the LLRV used a "collective" control, similar to a helicopter, the LLTV used a lift rocket throttle. Casters on the legs of the first LLRV would be replaced by pads to stop unwanted sideways movements at take-off and landing.

Artist's impression of the LLRV in 1963. Note the glass enclosure proposed for the pilot. This would be replaced on the LLTV by a cockpit with reduced visibility to more accurately simulate the conditions in the LM.

[Grumman LEM TM-1 1964]

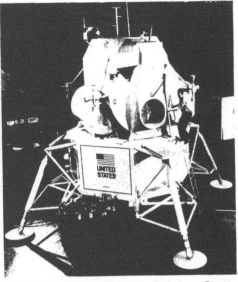

LATEST LEM MOCKUP—The Apollo TM-1 (Lunar Excursion Module) is displayed at the Grumman Aircraft Corp., Bethpage, N.Y. A Grumman engineer climbs the ladder to the entry platform.

Just a week before the arrival at Dryden of the first LLRV, *Grumman* publicly revealed the next iteration of the Lunar Module. The latest "mass report" had been issued on March 1st and for the first time it showed the non-symmetrical LM, but with rounded propellant tanks. It also appeared to have a narrower spread on the deployed landing legs. This one became known as LM TM-1 and was built full scale, out of wood. It was unveiled on March 24th 1964.

Later that Spring consideration was finally given to how an astronaut would get down to the lunar surface. Since the hydraulic egress platform had been left behind the previous summer, at first a rope ladder or even a knotted rope was considered, although quickly disregarded as impractical, the two photographs below are quite revealing in that they show a ladder attached to the descent stage wall which is then subsequently removed in favor of trying the rope. This in itself tells a story of weight and design problems. The ladder's position is in a most inconvenient location, mounted right on Quad 1. If it had been left as illustrated, it would have compromised later decisions to store the S-band erectable antenna and later the lunar rover in that same location. Clearly the best choice was to mount the ladder right on the forward leg. Even though *Boeing* were no longer anywhere near the LM pro-

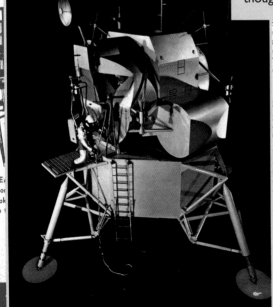

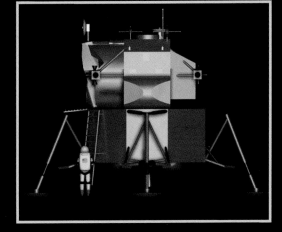

gram, artist impressions of the *Boeing* LEM suddenly appeared around this time showing a ladder on one of the legs. (see page 52)

At about the same time as the ladder issue was being resolved it had been finally decided to give up on the idea of having two docking hatches. By removing one it saved weight and allowed for a much simpler egress hatch to be added. The front hatch was to have been used for a LM-controlled docking with the Command/Service module. Of course by removing this front hatch it would mean that the LEM pilot would no longer be able to see the command/service module during docking. The solution to this was to add a small window in the top of the LM cabin. This would allow the pilot to look straight up above his head and see the approaching mother-ship during docking.

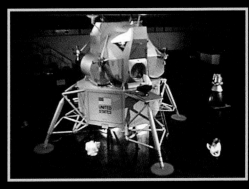

LEM TM-1 is studied at Grumman (above) and a similar mockup at the 1964 World's Fair (below)

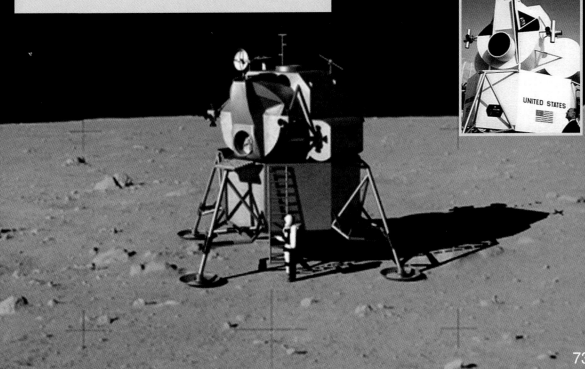

[Grumman LEM M5 1964]

LEM TM-1 finally brought some closure to the large-scale structure of the lander. The design would be more-or-less locked-in, and by the Fall it would be rendered in aluminum. Between October 5th and October 8th 1964 *Grumman's* metal LM mock-up, the M-5, was inspected by a review board which included Robert Gilruth, Joe Shea and Owen Maynard. This meeting took place at the *Grumman* factory in Bethpage New York and was the last major opportunity to propose fundamental changes to the LM configuration. Astronaut Jim Lovell climbed into a test space suit and was attached to the "Peter Pan" rig that simulated lunar gravity, while the committee and the press observed. Meanwhile, *Grumman* pilot Robert Smyth erected the large orange S-Band antenna. At this time the stowage position for the S-Band was still not finalized. The two front Quads of the M-5 had metal panels with hemispherical bulges, presumably for internal tankage.

The metal *M-5* looked very similar to the final LM and from this point forward most of the changes were internal, with the major exception of changing the shape of the front hatch from round to roughly square in January of 1965. Just a week earlier the Command Module mock-up had been reviewed at the *North American Aviation* factory in Downey California. A press conference was held at this event at which astronauts Deke Slayton, Alan Shepard and Pete Conrad discussed their opinions of the new lunar spacecraft.

BIRTHDAY PRESENT? — Astronaut Russell L. Schweickart, on his 29th birthday, receives a packet of dehydrated pea soup from Herbert Greider, test conductor responsible for a week-long test of biomedical equipment at the NASA Manned Spacecraft Center. Helping him celebrate his birthday are, from left, Greider; H Vick, a bioinstrumentation specialist; Tom Turner, a Gemini suit specialist, and Hal Parker of the Flight S tion Branch. The cake was brought to Schweickart on Saturday night by his wife Clare. She also presente with a toy electric razor that "sounds like dad's" but doesn't shave, and a bar of deodorant soap. (Schwe was unable to bathe during the eight-day test.) Schweickart, who was testing freeze dehydrated space on Sunday, watched as members of the test team devoured his cake.

LEM Checkout Conducted By NASA At Grumman

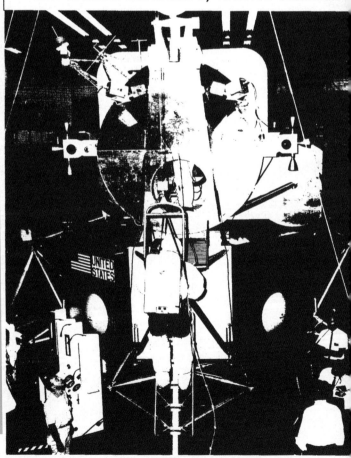

S-BAND ANTENNA — Robert K. Smyth, LEM consulting pilot for Grumman Aircraft, demonstrates the S-band antenna for a press conference at Grumman, Bethpage, N. Y., October 8. The antenna can be stowed in a cylinder 10 inches by three feet. It unfurls like an umbrella and can be set up on the lunar surface to aid TV transmission to earth.

LEM COCKPIT CHECKOUT — Shrouded in a thermal suit over his regular pressure suit, with operable back Astronaut James Lovell prepares to check LEM cockpit for accessibility during the NASA Inspection and R of the all metal mockup of the Lunar Excursion Module, at Grumman, Bethpage, N. Y., October 5-9. wearing the "Peter Pan" hoist rig that simulates the one-sixth gravity condition on the moon.

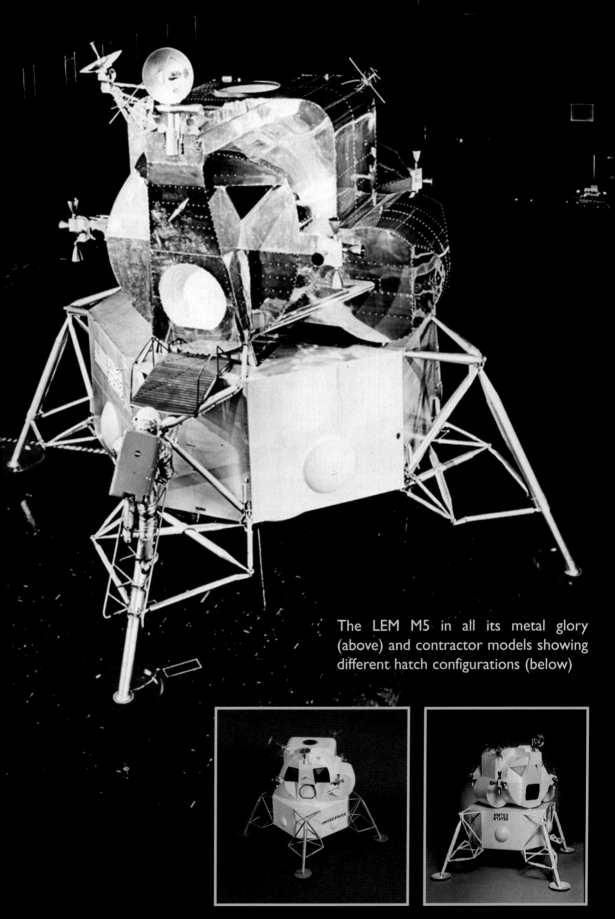

The LEM M5 in all its metal glory (above) and contractor models showing different hatch configurations (below)

[Extended LM 1966]

On January 22nd 1968 the first lunar module would be launched into space aboard a *Saturn IB* launch vehicle. It was dubbed *Apollo 5* and the major distinction between the *Apollo 5* LM and the final LM was that it had no legs, since they were not deemed necessary for the mission objectives. It would fly unmanned in low earth orbit before burning up in the atmosphere.

The next big thing for *Grumman* was to secure a future in which it could take all that it had learned in designing the LEM and turn it into an assembly line for future missions. This would involve taking rovers, habitats and flying vehicles to the lunar surface aboard the LM descent stage, meanwhile the ascent stage could be adapted for use as orbital habitats for telescopes, mobile rovers and laboratories.

The proposed full-up versions of the LM included a basic "extension" package called, predictably *Extended LM*. This would make the assumption that after the first batch of landings, enough information would be learned to increase the payload without increasing the propellant load. This would enable the LM to stay on the moon for up to three days and carry an additional 700 pounds of equipment such as a small lunar rover or a lunar flying vehicle, such as the Grumman space-bike.

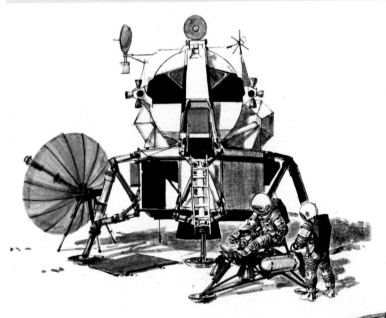

Grumman space-bike (above) superficially similar to the Marquardt Space Sled (top opposite), but with legs.

SPACE SLED

A design for individual space maneuvering in the vicinity of a space craft, the space sled lost out to the seat-like maneuvering unit see to the right. The mannequin riding the sled is wearing an experimental space suit which was one of a series tested in the mid-1960s. Unlike the space sled, the suit was part of a continuing evolution which produced the operational space suits used today.

[Apollo Applications Program 1966]

The next few pages cover the proposed modifications to the Lunar Module. In the mid 60s, NASA planned to use Apollo designed hardware to continue space exploration, these designs were all part of the *Apollo Applications Program* (AAP). The multifaceted program included Earth orbit (Skylab was the result) and lunar missions. One of the concepts studied was to modify Apollo LMs and CSMs to enable 14 day lunar missions. The modified LMs included the *Extended LM* (ELM), *Augmented LM* (ALM), *LM Shelter* (LMS), *LM Truck* (LMT), *Lunar Payload Module* (LPM) and *LM Taxi*. Combinations of these were also considered such as *ELM Taxi*, *ALM Taxi*, *ALM Shelter*, *Augmented LPM*, *Augmented LMT* (ALMT), *LMT Shelter* (LMTS) and *Augmented LMTS*. These different configurations could increase the LMs payload capacity from 300 lbs (LM) to 12,000 lbs (ALMT). The **ELM** was used for Apollos 15-17 and increased the usable payload by a factor of five over the conventional LM, to 1500 lbs. This was accomplished by not using a free-return trajectory, conducting lunar orbit separation at 50,000', having a range-free trajectory, no continuous abort, and optimising the launch time. Additional payload and hardware were attached to the outside. The **ALM** almost doubled the ELM payload to 2800 lbs. This was accomplished by redesigning the descent stage to accomodate larger fuel tanks, using an uprated Saturn V and adding an ablative coating to the LM descent engine nozzle. Both the ELM and the ALM could be flown unmanned which increased the usable payload by 600 lbs. They could also both be used as **LM Taxi** flights which meant they could be stored on the surface while the crew lived in another lander such as a LM Shelter. The **LM Shelter** was an unmanned lander, capable of being quiescent up to 90 days, housing two-men for 14 days, with 340 cubic ft of living space (88 for the airlock). It had no ascent engine, was powered by fuel cells, with cryogenic storage, radiators, provisions, bunks, 27 sq ft of floor area and could carry an LSSM. It could carry 5300 lbs of payload but using the augmentation package this could be increased to 7275 lbs. (ALMS). The **LM Truck** was a modified LM descent stage that came in two different basic versions the **Integrated LMT** and the **Modular LMT**. The ILMT had all of its guidance and controls (G&C) built into the modified descent stage while the MLMT had an additional G&C module attached to the top. The ILMT could carry 10,000 lbs and the MLMT 9800 lbs. By adding the augmentations (creating the ALMT) this could be increased to 12,000 lbs. A combination of LM Truck and LM Shelter, called the **LM Truck Shelter** could carry 8000 lbs payload and had a docking bridge allowing it to connect to the CSM. It also had a rigid airlock and a 1300 cubic foot inflatable cabin. The deflated cabin allowed for additional payload to be mounted on the top including an LSSM, a crane, a STEM shelter or other provisions. The floor area of the LMTS was 110 square feet as opposed to about 14 square feet in the standard LM. By adding the augmentations (ALMTS) payload could be increased to 10,000 lbs. The **Lunar Payload Module** (LPM) was an unmanned LM stripped of all life support equipment, human accomodations and controls with no ascent engine. It was capable of staying quiescent for up to 90 days, could carry 8400 lbs of payload in a volume of 730 cubic feet. It could carry an LSSM. With the augmentation package (ALPM) it could carry 10,400 lbs. The ALMT could carry a MOLAB or MOBEX up to 12,000 lbs, but the heavier versions of MOLAB required a custom LLV (see page 145).

[LM Taxi 1966]

The *LM Taxi* differed from its parent spacecraft due to a few subtle modifications necessitated by the slightly different nature of its mission. The *LM Taxi* mission would ferry two astronauts to the moon on the second leg of a dual launch following the successful landing of a logistics vehicle on the first leg. This logistics vehicle would usually be a *LM Truck* or *LM Shelter* (although at least two larger versions would use a custom LLV). The *Taxi* had the same life support provisions as the Apollo LM. The astronauts would shut down and store the *Taxi* after landing and transfer quarters to the logistics vehicle for the 14 day stay. However, if the men found the logistics vehicle unsatisfactory for habitation, a contingency mission would be performed by having the astronauts return to the *Taxi* and live there for the duration of the mission. To do this they would first unload the modules of expendable supplies from the logistic vehicle and transport them the short distance to the *Taxi* using a lunar rover offloaded from the logistic vehicle. Umbilical from the modules were connected to an external panel on the *Taxi*, transfusing life-support fluids to the spacecraft and enabling the astronauts to perform the alternate mission. If the logistics vehicle did not include living quarters, this cargo transfer and *Taxi* activation would be the normal procedure for a 14 day mission. The *Taxi's* return mission profile was a duplicate of the Apollo LM. However, ascent payload could be increased in two ways; either by topping off the ascent tanks prior to Earth launch, or by reducing the lunar ascent propulsion budget.

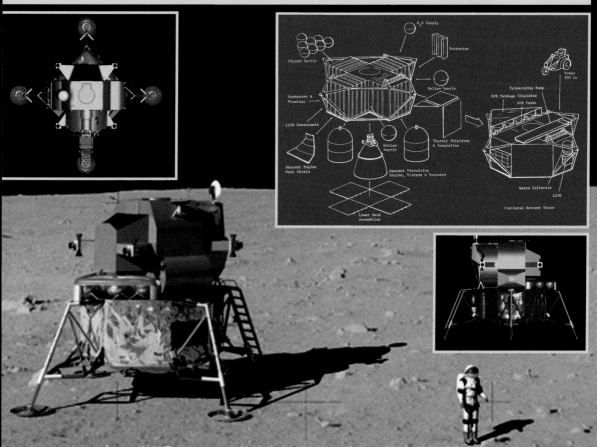

Grumman artist's impression of LM Taxi (opposite) Also note extra array of oxygen tanks behind the ascent stage in the rendering above. This optional feature was possible because one configuration of the Taxi had an empty area on the back for just such a purpose. In the diagram above you can see the proposed location of the ramp and mini rover which would be placed next to the exit hatch. This exact configuration migrated to the LM Shelter concept.

[LM Shelter/LPM 1966]

The *LM Shelter* was an Apollo LM minus its ascent propulsion system, but modified to make an unmanned landing on the moon, remain quiescent for up to two months, and support two men for 14 days. A *LM Shelter* launch and successful landing would be followed by a manned *Taxi*, in a dual launch mission. *Shelter* payload would consist of expendables, mobility aids (such as a rover), a 30 meter lunar drill and an advanced ALSEP. It would also have carried along a large solar array that the astronauts could deploy to provide additional power to the *Shelter*. This proposal from 1966 for the *LM Shelter* still retained the basic structure of an ascent stage, but without the engine or propellant. Strapped to the side was one of three different sized rovers and a scientific payload. A *LM Shelter* without life support, controls, human accomodations or ascent capability was known as the *Lunar Payload Module (LPM)*. It was capable of delivering over four tons to the surface which could be increased to five if it used a series of proposed augmentations.

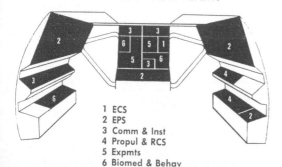

This artist's impression (below) shows version 3 of the proposed LM Shelter with a shelf life of 3 months and mission duration of 14 days. Powered by *Gemini* fuel cells and equipped with a medium range rover, all weighing in at **6944** pounds. At left are the proposed control arrangements and, below that, two different configurations that had an airlock attached to the nose of the shelter.

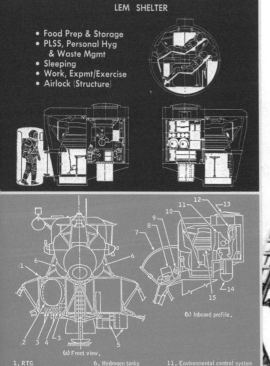

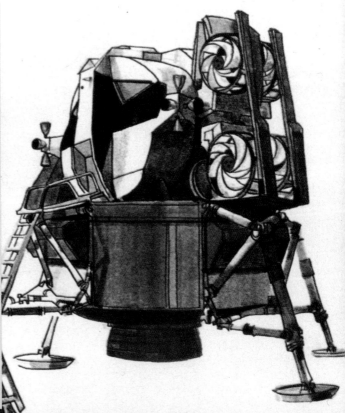

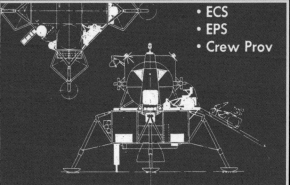

- ECS
- EPS
- Crew Prov

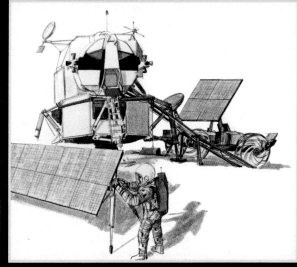

SHORT-RANGE ROVER

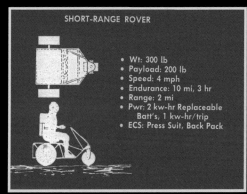

- Wt: 300 lb
- Payload: 200 lb
- Speed: 4 mph
- Endurance: 10 mi, 3 hr
- Range: 2 mi
- Pwr: 2 kw-hr Replaceable Batt's, 1 kw-hr/trip
- ECS: Press Suit, Back Pack

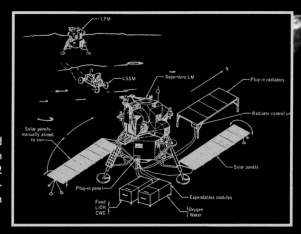

This artist's impression at right shows an LPM and LM and the proposed layout of the delivered equipment to sustain the LM for 14 days. The renderings below show version 2 with a medium rover and a medium payload on the opposite flank. Above is the short range rover and a diagram showing where it was to be stored during transit.

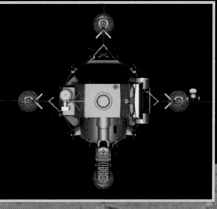

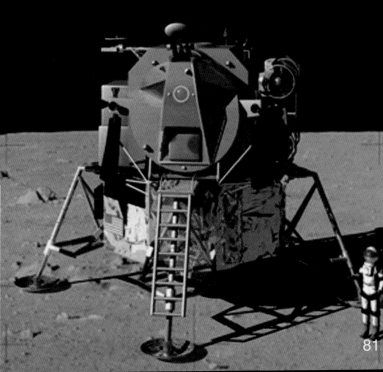

81

[LM Truck/LM Truck Shelter 1966]

The *LM Truck* was an unmanned lunar lander that transported cargo in the volume otherwise occupied by the LM ascent stage. The payload fairing was essentially a frustrum of a cone with an 18 foot diameter base, that was 10 feet high and with a top diameter of 15 feet. Components from the removed ascent stage still vital to the *Truck* mission were relocated to a central docking structure attached to the existing interstage fittings on an unmodified descent stage. The docking structure enabled the CSM to transpose and dock to the *Truck* and extract the *Truck* from the Spacecraft LM Adapter (SLA). Thereupon an astronaut from the CSM could reach into the structure through the docking tunnel and, using a keyboard stored in the structure, update the *Truck's* guidance, navigation and control subsystem so that the spacecraft's trajectory to landing was held within acceptable tolerances. A *Truck* landing represented only half of a dual launch mission. Once the payload arrived successfully on the moon, a second Earth launch of a *Taxi* dispatched astronauts who then used the life-support and scientific equipment on the *Truck* in supporting their duties during long-term lunar explorations. A typical mission, shown in this artist's rendering, might have comprised a Lunar Roving vehicle, resupply modules for supporting two men in the *Taxi* up to 14 days, and a 3700 pound, 400 cubic foot scientific payload. Several versions of the *LM Truck* included a crane to facilitate the off-loading of the payload.

This artist's impression shows the *LM Truck Shelter* (LMTS). The docking bridge can be seen in the center. The cabin on this version was inflatable so that more cargo could be carried. This shows an LSSM being offloaded by a crane. In this case the ILMT has been used for the descent stage. NASA expected it would take 42 months to construct the LMTS and expected it to be ready by 1971.

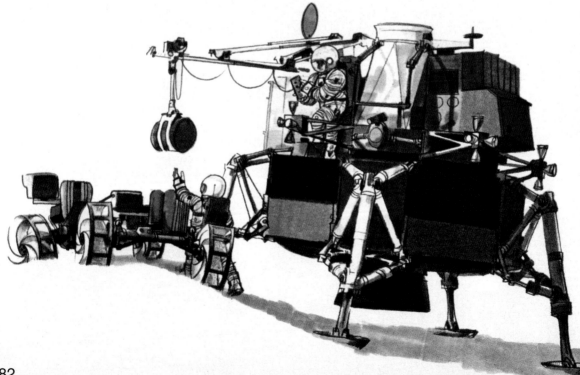

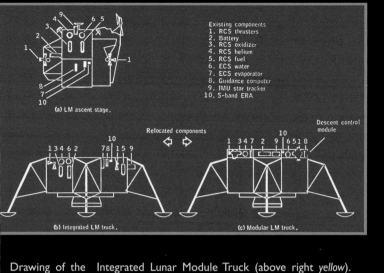
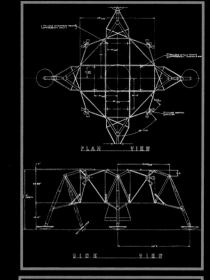

Drawing of the Integrated Lunar Module Truck (above right *yellow*). Shows one proposed configuration for the RCS system on the descent stage. The rendering below demonstrates the hold-down architecture for a rover and a flying vehicle. A truncated cone shaped volume could basically be used to fly almost anything in place of the conventional ascent stage. Another version called the MLMT (Modular Lunar Module Truck) had docking, guidance and control modules on top of the descent stage to provide an unmanned logistics payload lander (see comparison above). Rare ILMT contractor model (bottom right)

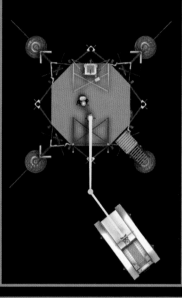
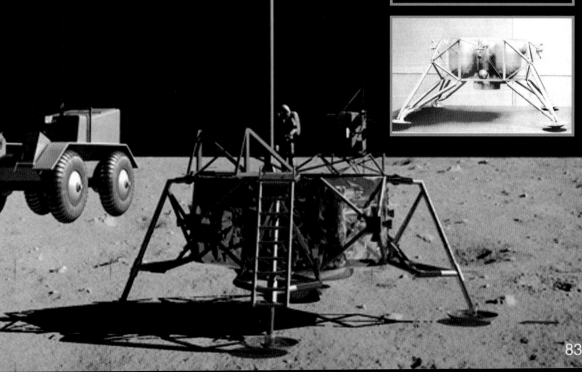

[Goodyear STEM 1965]

In 1965 Goodyear corporation of Akron Ohio submitted their proposal in response to a contract issued by NASA Langley for an inflatable lunar shelter. This Stay Time Extension Module (STEM) was a two-man system designed for 8 to 30 days of operation on the lunar surface. It featured an airlock made of pliable and expandable material and could be erected by a single crewman. The final inflated length was 210" with a height/diameter of 84". Inside were two bunks, a toilet and a worktable. The entire package was built in full-scale mock-up by Goodyear and the packaging design allowed it to be flown strapped to the quads of a regular LM as seen in the deployment sequence below. Later it could have been flown aboard a LM Truck Shelter (right) like the one seen on page 82.

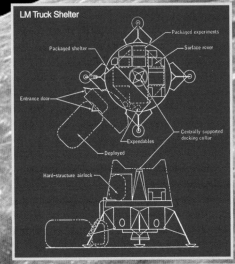

LM Truck Shelter

The full-scale STEM is seen here in 1965 and the internal design is shown below.

STEM weighed 1600 lbs and would be loaded in a canister on the side of a LM Truck or ALMT.

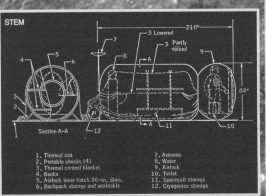

STEM
1. Thermal mat
2. Portable chocks (4)
3. Thermal control blanket
4. Bunks
5. Airlock inner hatch 36-in. diam.
6. Backpack storage and worktable
7. Antenna
8. Water
9. Airlock
10. Toilet
11. Spacesuit storage
12. Cryogenics storage

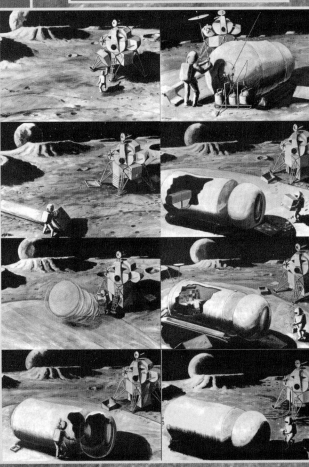

[NAR CM Shelter 1965]

Also placed under serious consideration was the notion of putting a heavily modified Apollo Command Module atop a LM Truck. The original idea for this came from NA Rockwell engineer James Matzenauer. His final patent showed a standard LM descent stage with the RCS system mounted on the CM. Originally plans used the ILMT. This CM Shelter would have had considerably more living space than the traditional LM ascent stage but of course would have been solely usable as a short-term habitat since it would have had no ascent engine.

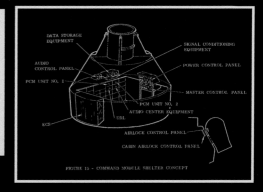

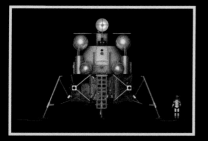

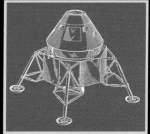

Matzenauer's original patent drawing for CM shelter (right blue)

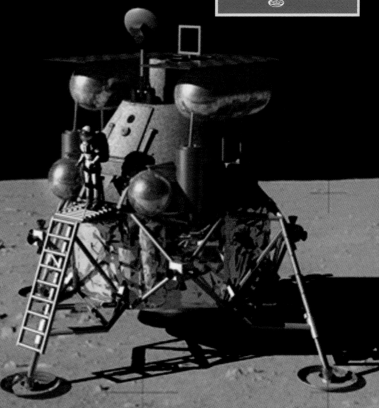

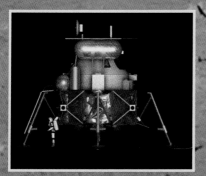

[SHELAB 1964]

An extremely sophisticated proposal for the *LM Truck* became known as the *SHELAB*. This version was devised by the *Hayes International Corporation* under a November 1964 contract issued by NASA in Huntsville. Considering how soon after the final decision was made on the design for LM M-5, it is remarkable how fast *Hayes* pounced on the possible uses for a reconfigured LM. The *SHELAB* was defined as a two-man shelter/laboratory (thus *SHELAB*). Mounted atop the LM Truck were a large volume habitat, an airlock, a lunar roving vehicle and a sizeable storage area of consumables. Once *SHELAB* landed on the moon it was there to stay, since it had no ascent stage. This habitat was designed to support a crew of two for up to 14 days and included a reasonably large living area (about 16 feet in diameter and almost ten feet high) that included a kitchen and two bunks slung high up inside the pressurized cabin. The airlock served a dual purpose as a solar flare shelter. It was to have a meteor bumper enclosing a polyurethane foam lining, some "super" insulation and an inner aluminum skin. The airlock sat at one axis of the center of gravity of *SHELAB* and was designed to have a docking adapter on the top to facilitate the retraction from the SLA after launch.

The whole cabin and airlock was mounted on a large ball joint with two hydraulic jacks at the rear, to level the habitat in case the *LM Truck* landed on uneven ground. The total combined weight of the *SHELAB* package was 6500 pounds and was known as the ALSS module. Another unique thing about *SHELAB* was the decision to equip it with a 1500 pound lunar roving vehicle that carried its own onboard science lab, including a large drill mounted down through the center of the vehicle. It also had individually powered four wheel drive, with wheels that were made of flexible wire. Its unloading mechanism was given considerable thought and bears more than a passing resemblance to the *Boeing LRV* system that flew on Apollo 15 in 1971. This rover might seem a bit large and bulky compared to what actually flew in 1971 but the driver would also have access to his own personal lunar flying device, a strap-on rocket belt. The rocket belt had a range of 2 miles and was incorporated into the rover in case of emergencies, or if the rover was incapacitated.

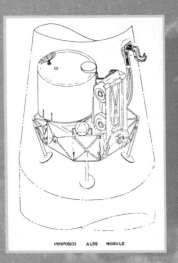

SHELAB inside the Saturn LM housing.

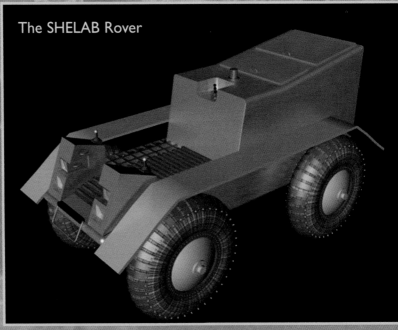

The SHELAB Rover

Cutaway showing the inside of SHELAB

[AiResearch ELS 1967]

The *LM Truck* was also considered for a habitation module by *Garrett AiResearch*. This was known as the *Early Lunar Shelter* (ELS). The primary objective was the evaluation and conceptual design of two and three-man lunar shelters for comparison with competing concepts such as the *LM Shelter*. The minimum design had a pressurized volume of 500 cubic feet for two-men on a 14 day mission. Final designs were 750 cubic feet. Basic shelter equipment included rechargeable PLSS, fuel cells for EPS, cryogenic supercritical storage for fuel cell reactants and life support gases and Lithium hydroxide for CO2 removal. The total package weighed 10,300 pounds including 1345 pounds for the structure. *Grumman* artists showed the *AiResearch* habitation module equipped with a *Grumman* rover. At least four different designs existed for the ELS, the one shown opposite is Configuration 3B using a horizontal cylinder. Version 4 used a vertical cylinder for the crew compartment (diagram opposite).

Artist's impression of an ELS. This one shows a redundant ladder with side hatch and the rover being off-loaded using a ramp. Later versions placed the hatch at the end of the compartment (see the interior of configuration 3B at right) and used an A-frame pulley and winch system to off-load the rover. Lander would be an LMT or ALMT. Later versions may have been powered by a SNAP-27 RTG.

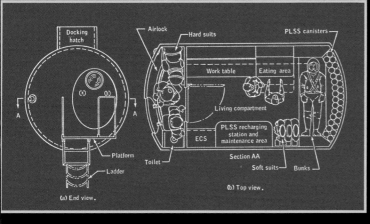
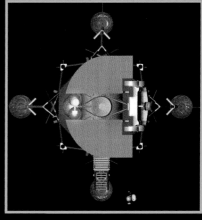

Diagram showing proposed interior of another ELS (above)

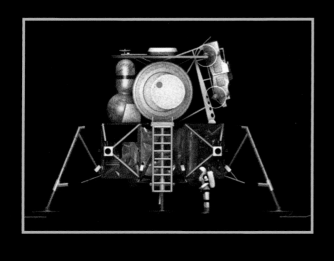
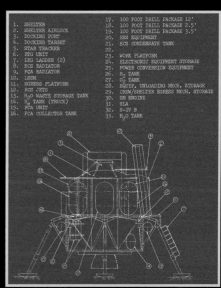
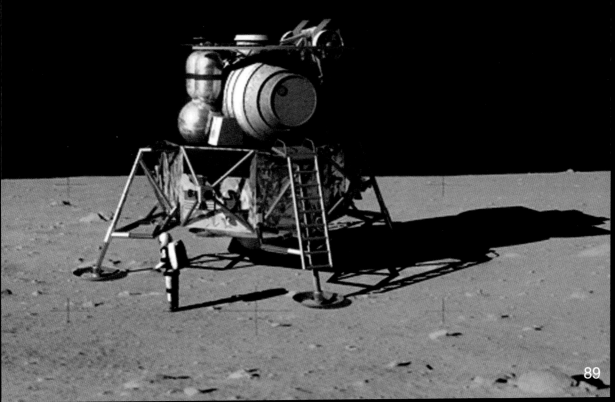

[LM Lab 378B 1966]

Other functions were considered for the LM including as a rescue vehicle, a stellar observatory and a lunar orbiting laboratory. The orbiting laboratory employed highly modified ascent and descent stages. The legs were removed and a bank of instruments, telescopes and even small rocket launched probes were mounted around the two stages. Many different versions of the LM Lab were proposed by *Grumman* under their Design 378B program. All of these LM labs would have been manufactured as LEM spacecraft at *Grumman* and then modified at KSC. The burden of modification would be mostly on the experimenters and their experiment packages rather than on the contractor. Model 507 would have been launched into a polar orbit with an accompanying CSM by a Saturn V. This would have been a preparation flight for lunar surveying. The mission would last 11 days. Model 509 would have been in an equatorial orbit, again accompanied by a CSM. This would have been primarily concerned with astronomy in the X-Ray, UV and radio frequencies. It would also analyze the Martian atmosphere. Model 214 would have been flown aloft by a Saturn IB and would have rendezvoused with a separate CSM. This mission was to study rendezvous techniques but also was to carry enough consumables to act as a 24 day space station for longer duration medical studies. Model 216 would have been another dual Saturn IB launch and would have stayed in low Earth orbit for 20 days to study EVA techniques and medical responses. The one pictured here was designated Model 511 and would have been used in lunar orbit for about five days. It would have been launched with a CSM aboard a Saturn V. The main goal would have been studying lunar geology using a broad array of sensors. It would also have concentrated on finding suitable landing sites and accurately identifying their location.

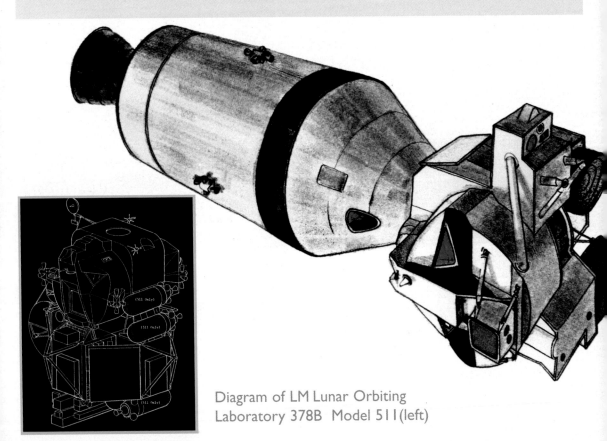

Diagram of LM Lunar Orbiting Laboratory 378B Model 511 (left)

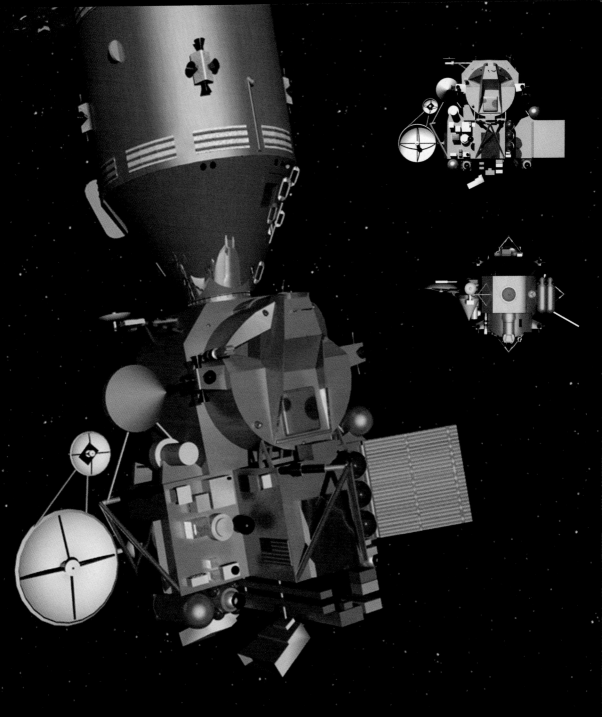

LEM LAB

- Food Prep & Storage
- PLSS, Personal Hyg & Waste Mgmt
- Work, Expmt, Exercise

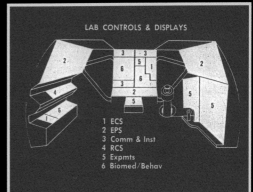

LAB CONTROLS & DISPLAYS

1 ECS
2 EPS
3 Comm & Inst
4 RCS
5 Expmts
6 Biomed/Behav

[Rescue LM 1966]

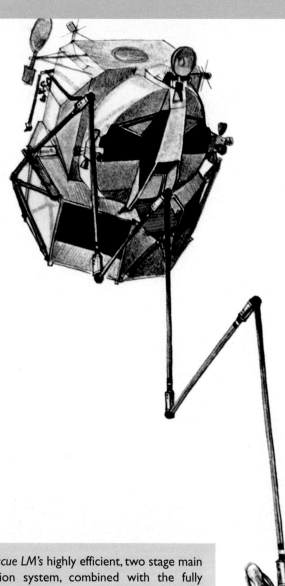

The *Rescue LM's* highly efficient, two stage main propulsion system, combined with the fully redundant reaction control system and versatile guidance and navigation capability, offered an in-orbit maneuvering, manned vehicle of unique capability. The application of this capability by a LM with very limited modifications would permit in-orbit intercept of, and rendezvous with, other space vehicles for such purposes as personnel rescue (as depicted here) and spacecraft inspection and repair.

[LM ATM 1966]

The *LM/Stellar ATM* was to evaluate performance essential to the development of advanced solar and stellar observation systems. The experience gained on the *LM/ATM* could have been applied to other observatory missions by replacing the solar telescope with a large aperture stellar telescope. Such a configuration could be operated in a manned or unmanned mode of operation.

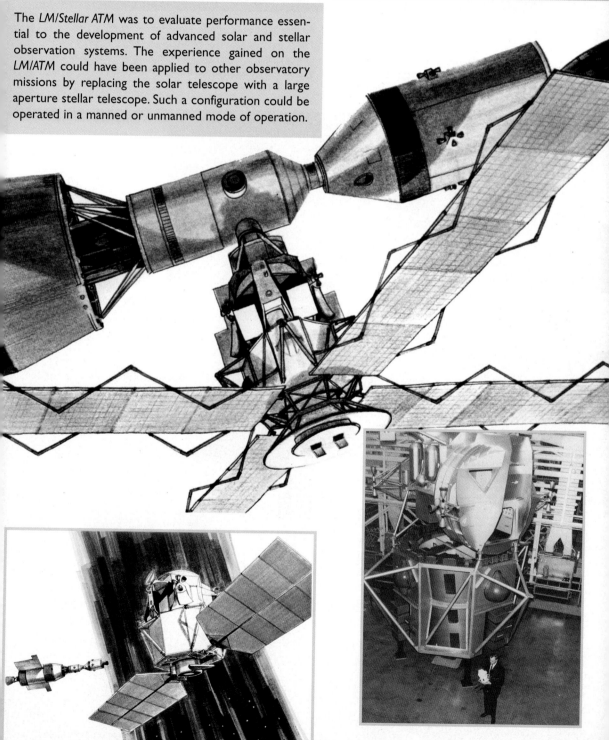

[Grumman LM 1969-1972]

The Grumman LM finally flew in space in March of 1969 for the flight of Apollo 9. Some minor modifications were still made before the first moon landing. Such as the addition of reaction control deflectors, a series of metal plates designed to divert the exhaust away from the delicate walls of the descent stage. The final training vehicle is seen below, while opposite is a fully restored flight vehicle on display at the National Air & Space Museum in Washington DC. Inset at right is a Grumman desk model.

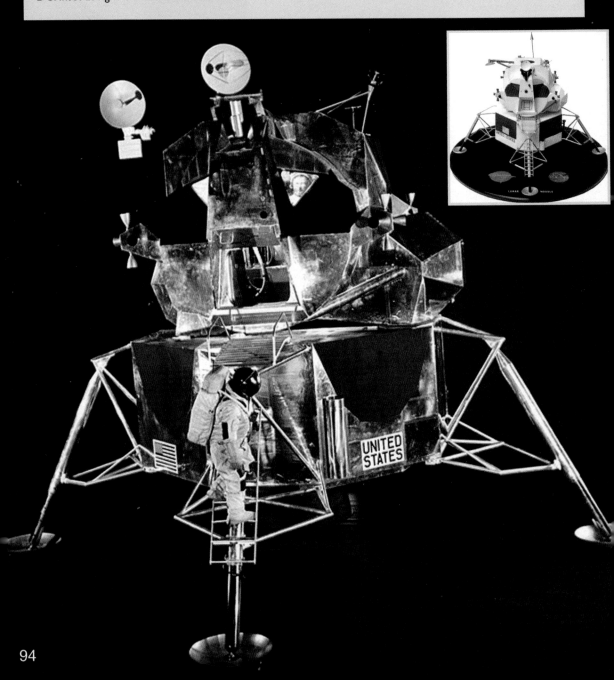

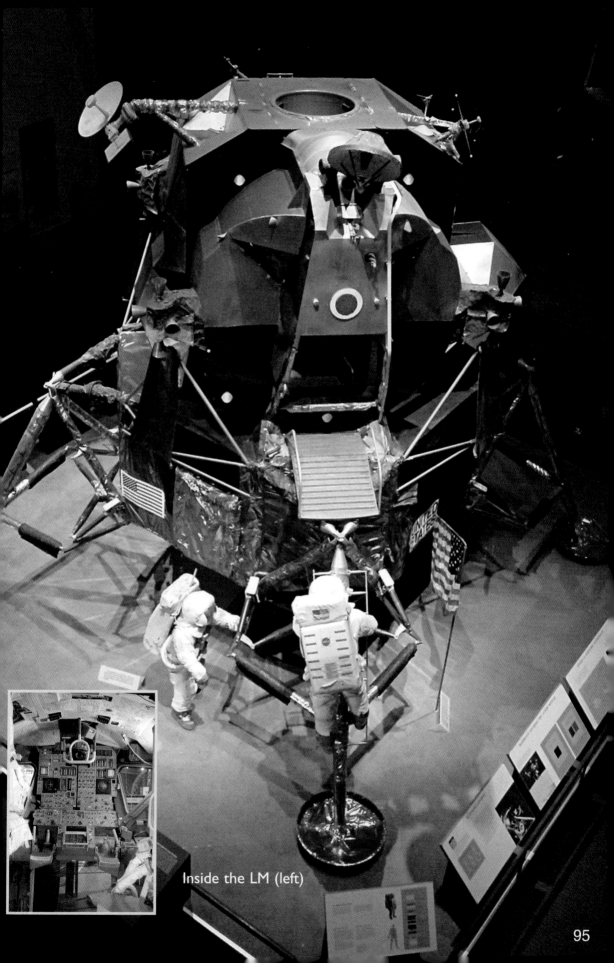

Inside the LM (left)

[LM Controls 1969-72]

Some slight modifications were made to the Lunar Module flight controls between 1969 and 1971. This diagram shows the basic layout (explained in the legend at top right). The commander stood at the left side of the vehicle and performed the landing while the Lunar Module Pilot monitored position, rate of descent, fuel consumption, altitude and the main computer from the right console.

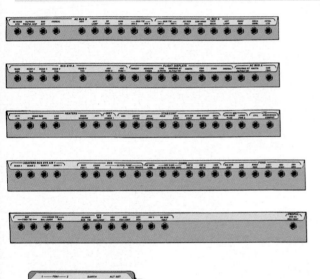
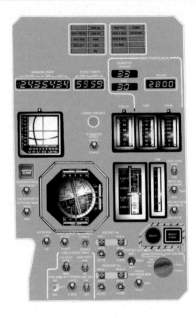

UTILITY LIGHT

ALIGNMENT O

ORBITAL RATE DISPLAY — EARTH AND LUNAR (ORDEAL)

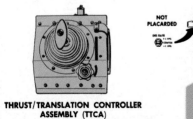

THRUST/TRANSLATION CONTROLLER ASSEMBLY (TTCA)

ATTITUDE CONTROLLER ASSEMBLY (ACA)

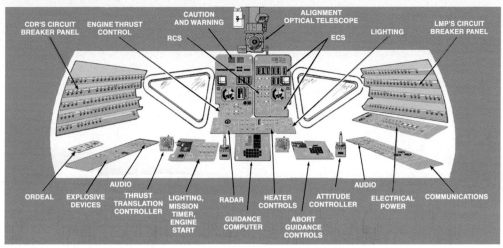

Control and Display Panels Locator

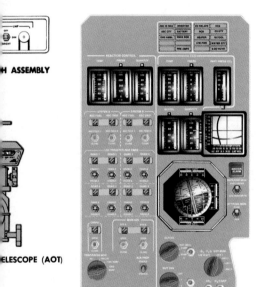
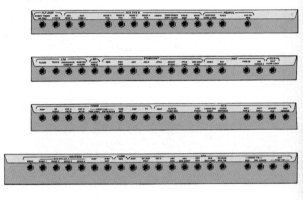

THRUST/TRANSLATION CONTROLLER ASSEMBLY (TTCA)

ATTITUDE CONTROLLER ASSEMBLY (ACA)

[Chrysler LRV 1963]

Chrysler Corporation, working in consort with the Society of Automotive Engineers, produced a comprehensive study in 1963 of the requirements for a lunar surface rover. Their conclusions favored either tracks or wheels over more exotic proposals such as hopping or walking machines. The Chrysler LRV was to carry two crew in a pressurized environment at up to 10 mph over a range of 80 miles but had a maximum range of 200 miles. It was to be delivered in pairs to the lunar surface by a Saturn V carrying a Chrysler Lunar Freight Vehicle. It was able to withstand at least one major solar event since the shielding was over half of the 6000 pound weight of the vehicle. The crew exited via large hatches in the front which also included portholes. A 360 degree periscope with a floodlight attached was placed on top; alongside antenna for communications with Earth. The five foot diameter wheels doubled as radiators and provided 18 inches of ground clearance. The tracked version used slightly smaller wheels driven by rear mounted electric motors.

[LESA] 1963-1965

The Boeing company filed a response to a Marshall Space Flight Centre RFP and won a contract to study lunar habitats. In March of 1964 Boeing filed its report called *Lunar Exploration Systems for Apollo* (LESA) and continued to supplement that report well into the third quarter of the year. Boeing wanted to demonstrate a system that was modular and expandable to provide a long term lunar architecture. LESA was designed to cover all possible parameters from the specified 90 day stay of a two man crew up to a possible two year system housing a population of 18 crew. In February of 1965 both Lockheed and GE put forward their own versions of LESA (see diagram).

The proposed Boeing LESA system called for a family of prefabricated modules that could be assembled on the lunar surface in an assortment of configurations to support different missions. All of the proposed modules would have been taken to the moon atop an appropriate LLV aboard the Saturn V. A maximum diameter of 260 inches, with a maximum weight of 25,000 pounds was allowed for the various modules. Once on the surface the LESA habitats would undergo some startling transformations. Perhaps the most intriguing was the idea of using a caisson to provide long term radiation shielding. This peculiar system involved the expansion of an extra aluminum wall around the main crew quarters and then filling it with lunar regolith. Apparently this was considerably cheaper than trying to either fly adequate shielding from the Earth or burying the habitat. Of course the assembly and maintenance of such a facility would require an extensive array of mobile vehicles. In July of 1964 Boeing submitted their final report for a family of possible roving vehicles to support LESA. This study comprised six different sections-broken down as follows: Basic Rover - Unitized, Basic Rover - Modular, Intermediate Rover - Unitized, Intermediate Rover - Modular, Special Vehicle Concepts, Flight Vehicle Concepts.

MOVING CONCEPT
APPROXIMATELY 4000 POUNDS LOAD PLACED ON LRV TO INCREASE TRACTIVE EFFORT

LESA was also mobile

LESA SHELTER INBOARD

[Boeing Unitized Local Rover 1964]

The LESA Basic Unitized Local Rover was 10 feet across the wheels with a cabin providing 250 cubic feet of space. It had a 108 inch wheelbase with 70 inch wheels. A full airlock at the rear of the cabin also featured a toilet facility. A single bunk stretched across the right side of the cab and two swivel seats were provided for the operators. In one version it was considered that the "garage" door from the habitat module would double as the radiators for the rover. A small RTG was carried in the rear service section along with, batteries, water, scientific equipment, LH2 and LO2 supplies. The main cabin had facilities to store extra clothing and food as well as the main air conditioning system. Two TV cameras protruded from the side of the cabin as well as extra radiators and lights. A "fifth wheel" steering mechanism was installed between the two modules. Total weight was 4950 lbs. An almost identical modular version used a detachable chassis that allowed for the cabin and service section to be lifted off and replaced with other modules or cargo. The so-called Basic Modular Local Rover had 80 inch wheels and a 120 inch wheelbase to accommodate larger payloads. It weighed 5440 lbs.

[Boeing SCV 1964]

The Boeing Specialized Construction Vehicle (SCV) came with a 110 inch wheelbase, 90 inch diameter rear wheels, a 200 cubic foot cabin capable of carrying two crew, a fifth wheel assembly for towing habitat modules and cargo and a cutting auger with gravel throwing unit mounted on the front. What seems somewhat outlandish was the proposal to use the gravel thrower to literally toss massive amounts of regolith over 25 feet high and into the top of the habitat caisson. Thus filling the caisson with soil to act as a radiation shield. Presumably there was some plan to make sure the gravel thrower didn't damage the side of the habitat while this process occurred. On a more practical note the auger and thrower could be used to excavate a site to bury the LESA nuclear power plant, as long as the breccia wasn't too cohesive. The gravel thrower conveyor belt would operate at a 55° angle and at 26 fps. The helical cutters would feed the central helix, which in turn fed the thrower.

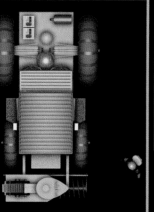
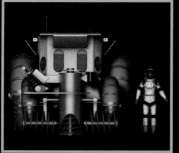

[Boeing 2 Man Intermediate Recon 1964]

The smaller two-man intermediate reconnaissance rover had an expanded cabin encompassing 325 cubic feet of space while still retaining the large airlock and toilet facility seen in the basic rover. The service trailer was also enlarged to provide more consumables and an equipment storage was also added to the back of the trailer. Total mass of this configuration was 7570 pounds with 800 pounds of that used for radiation shielding and 849 pounds of cryogenics. The wheels were enlarged from 70 inches on the unitized local rover to 80 inches (like on the modular local rover). While the overall vehicle width was increased from 120 to 154 inches. This particular vehicle would be dispatched in pairs allowing four men to explore for up to 42 days. Both modular and unitized versions were considered.

[Boeing 4 Man Intermediate Recon 1964]

Designed to be available as both a unitized or modular vehicle (with slight differences) the 4 man intermediate reconnaissance rover was to be deployed in pairs, each housing 2 men, but both with extended supplies for long duration missions. The cabin was expanded to 390 cubic feet by placing the air conditioning equipment over the right front wheel and the food and clothing storage over the left front wheel. Two bunks were placed along the right side of the cabin. The basic mission for this two rover configuration was 42 days of reconnaissance and exploration over a 600 mile range. The vehicle had a 67 cubic foot airlock and the main wheels were enlarged to 90 inch diameter. Because of the increased heat load the radiators were enlarged to include a 72 square foot roof radiator and two 35 square-foot radiators on each side of the cabin. The fuel cells could provide 926 kilowatt hours of total energy. Total mass of the vehicle was 9628 pounds and the entire package could still fit inside the 260 inch envelope provided by the Saturn V.

[Boeing Materials Handling Vehicle 1964]

Boeing's concept for a lunar materials handling vehicle was designed to be able to unload LLV platforms from a height of 185 inches and then transport those supplies to the crew shelter where it could raise them to a height of 280 inches. Boeing employed conventional forklift technology with the exception that the length of the lift exceeded normal forklifts. A single actuator provided the first 140 inches, this was then supplemented by a roller chain attached to the forklift slide to provide the remaining 140 inches. The vehicle had a cabin equipped for one man operation in a pressurized environment. Windows and driving controls faced in both directions so the operator could drive in either direction. A wheelbase of 140 inches allowed for a forward or backward tilt of up to 15 degrees. This was important since the cabin also rose with the forks to heights exceeding 18 feet. One set of wheels was on telescoping axles to provide additional balance. The telescoping axles also allowed for the Materials Handling Vehicle to be unloaded from its delivery craft through a smaller aperture. The entire fork and cabin assembly folded to a horizontal position to minimize the vertical profile of the vehicle during flight from 175 inches to 114 inches. Overall weight of the MHV was 3605 pounds. Back wheels were 90 inch diameter while the front wheels were 60 inch.

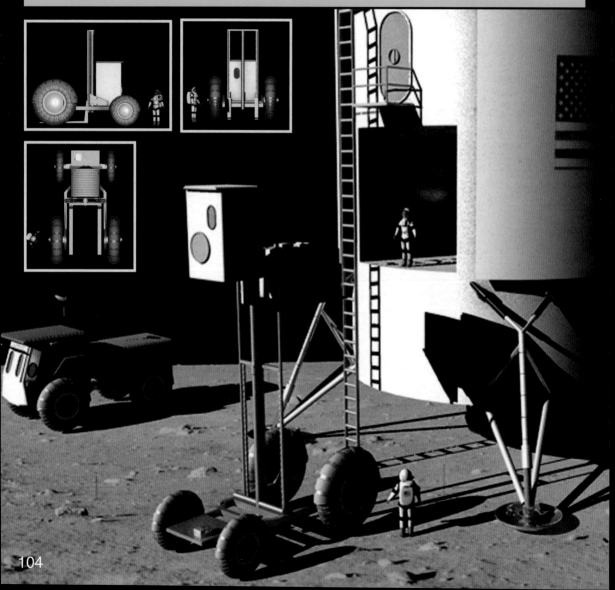

[Boeing 2-man Exploration Vehicle 1964]

The Boeing 2-man exploration vehicle was part of the LESA vehicle architecture and weighed 4550 pounds (terrestrial). It carried 1400 pounds of propellant and could carry 2470 pounds of payload. Round-trip range was 25 miles with 16 minutes of hover time. Delta-V was 5100 fps (2550 each way). The vehicle stood just over 13 feet high with a wide leg-span of 15 feet. A single main cryogenic engine provided 2500 lbs thrust fueled by pairs of LH2 and LO2 tanks. Twelve attitude control jets provided steering. Crew would pilot the vehicle in a standing position with "G" support provided by sling harnesses. Intriguingly, the vehicle could also be flown unmanned by remote control. Access to the cabin was through a top hatch, with a ladder provided for access. The cabin had no airlock so pressurization and depressurization protocols would have to be observed. Main electrical power was by fuel cell with batteries for auxiliary power. The landing gear included shock absorbers rated up to 6 g provided touchdown did not exceed 8 fps. Normal missions for the 2 man exploration vehicle would be 12 hours but supplies could be carried for emergency durations up to 48 hours.

[Boeing 1-man Hopper 1964]

The Boeing 1-man hopper was also part of the proposed LESA architecture. The hopper would be towed around on a launch trailer behind one of the many permutations of the LESA rovers. Its range was to be 10 statute miles. Gross weight was 1040 pounds with a payload capacity of 740 pounds and a hovver time of 10.5 minutes. The main purpose of this was to get to difficult-to-reach points of interest that the rovers could not manage. Like its two-man counterpart it was possible for it to fly by remote control in an unmanned mode. The equipment section, located behind the astronaut's back, contained batteries, two way communications equipment, and guidance and control electronics. The main cryogenic engine was fed from matching pairs of LH2 and LO2 tanks, while steering was effected by a reaction control system supplemented by main engine gimbals. The maximum range without a crewman aboard was 78 miles, making this an effective rescue device. In the case of a rover breakdown, the one-man hopper could be used to ferry crewmen back to the LESA habitat one at a time.

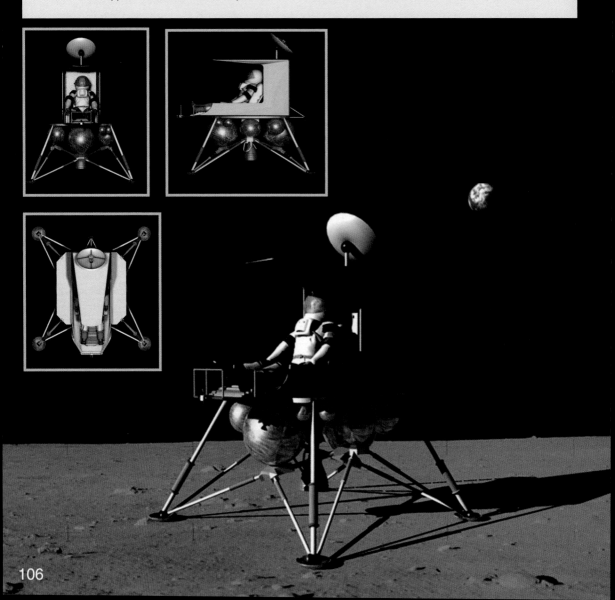

[Boeing FHI 1964]

The Flight Hover Indexer (FHI) was part of Boeing's 1964 Initial Concept for Lunar Exploration. The system illustrated below was conceived to be operable in either flight mode or surface-contact mode for maximum flexibility in unknown or difficult surface conditions or where the inherent time savings of flight between points could be decisive for human life or operational success. It was to be used for transporting personnel or critical equipment and could also have been used as a light work vehicle or a refueling post for personnel with rocket back packs. An unusual indexer protruding from the bottom appears to have acted also as a sort of shock absorber. The FHI could carry the Payload module (lower right) or drive around on the landing module. It could also be operated with two FHI in tandem for carrying larger payloads. The FHI was also capable of towing a trailer when attached to the landing module/rover.

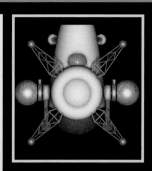

[GMDRL LSV 1964]

Boeing continued to study vehicles derived from the General Motors Defense Research Laboratory and included them in their Initial Concept for Lunar Exploration Study. This enormous Lunar Surface Vehicle (LSV) was to have weighed a staggering 24,000 pounds and its three separate modules were spread over 48 feet in length. The wheels would have stood 12 feet high and 2.5 feet thick. The cockpit stood over 16 feet off the ground. This extraordinarily optimistic design was to have been delivered in what appears to have been a custom landing craft that resembled the Saturn IV-B LASS concept. A large door would have opened in the side of the LASS and the LSV would have rolled down a ramp after having been flown in a vertical position.

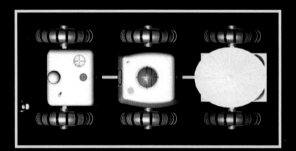
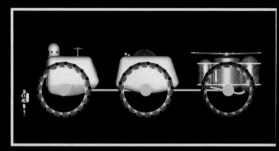

[GMDRL Cargo Hauler 1964]

Even more astonishing was the Maximum Concept Cargo Hauler from GMDRL. This gigantic cargo mover was to have stretched 85 feet from end to end with a 20 foot diameter wheel and a 24 foot wheel base. The whole machine would have been powered by either a SNAP reactor mounted in the rear axle or by fuel cells and tanks mounted at each wheel hub. Although the few surviving diagrams don't reveal much, it appears that this entire vehicle was capable of being coiled up almost like a telephone cable so that it occupied a footprint 20 feet high and 33 feet long. It was then to have been flown aboard a Saturn V. It was considered the maximum vehicle concept stowable in a Saturn V.

[Boeing V1 AMF 1964]

The Boeing V-1 AMF was the vehicle selected by Boeing in their 1964 study for supporting a two man crew for up to eight days. Three crew could be supported in the case of emergency. The vehicle was to be mobile over most lunar surfaces and capable of performing construction tasks using the construction module seen here. The backhoe could be replaced with an optional crane. The vehicle was 197 inches long, 108 inches wide and 104 inches high. It was to be powered by two 2-kw fuel cells and 5 kw-hr batteries. Four drive motors supplied four horsepower to the wheels. Cruising range was to be 300 miles. The backhoe was powered by a motor driven ball actuator. A second crew man could control the backhoe or crane from the standing position using a joystick, similar to the way that the Canadarm is controlled aboard the Space Shuttle. A robotic arm mounted on the front would allow for sampling without depressurizing the vehicle.

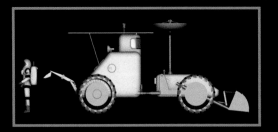
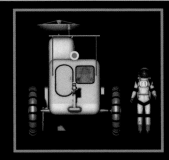
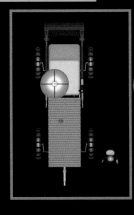

[Boeing V1 GMB 1964]

An alternative version of the Boeing V-1 was the GMB rover which came in both four and six wheel configurations. The extra two wheels were in the form of a front mounted module designated the V-2. This gave the rover an increased ability to cross obstacles during extended missions. It was mounted to the V-1 using a flexible chassis and this gave the primary rover the ability to climb over obstacles up to three times the wheel radius. An additional power supply of 2 kilowatts was carried within the module. The flexible couplings could be made rigid to improve the vehicle's ability to cross a crevice. The V-1 GMB could operate in four wheel mode without the V-2. This sleek looking vehicle had the same capabilities as the V-1 AMF and included a rear-mounted air lock.

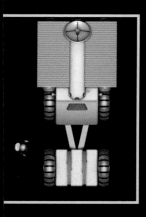
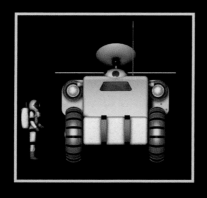

[Bendix 2 Wheel Cart 1968]

The Bendix two wheel cart was part of a 1968 presentation to NASA HQ called Lunar Surface Systems Technology. The cart could be used in either walking or riding mode. The long handles could act as a balancing skid when in riding mode. The cart was 124 inches long with a wheel span of 64 inches and a 32 inch wheel diameter. The forward storage deck could hold fourteen square feet of cargo. Steering was controlled by differential control of the speeds of the two powered wheels. Two control stations were provided with one at the end of the shaves and one on the riding station. The Bendix cart weighed 229 pounds and could carry 771 pounds over a range of 30 km at a top speed of 6 km/h. The Apollo Mobile Equipment Transporter used on Apollo 14 (bottom right) was similar in function if not in execution to the proposed Bendix cart. The MET had only one handle for towing and was 86 " long by 39" wide. It weighed 26 pounds and could carry 140 pounds. It actually used inflatable baked rubber tires that were filled with nitrogen at low pressure. Tools aboard included the lunar hand tool carrier and its geology tools, a closeup stereo camera, two 70 mm Hasselblad cameras, a 16 mm data acquisition camera with film cartridges, a sample bag dispenser, trenching tool, work table, sample weigh bags and the Lunar Portable Magnetometer.

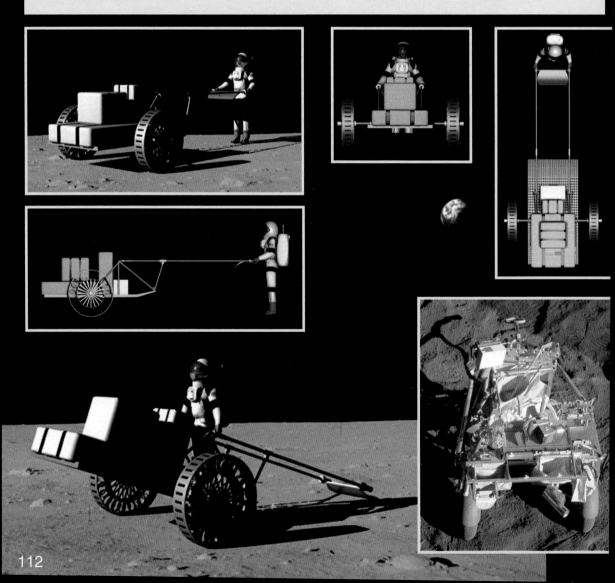

[Chrysler MLAV 1963]

Undoubtedly one of the strangest vehicles to come out of the early Apollo studies was the Chrysler Corporation's Manned Lunar Auxiliary Vehicle—a nuclear tricycle... The MLAV had a cargo tray 23" by 41" and could be operated with a hand controller either by a riding astronaut or by a remote control. A full size prototype was built and tested by Chrysler. The power source for this machine was to have been a SNAP 9A RTG, although it is not clear where this device would have been located. Other refinements would have included a television camera. The three wheels were tested with brushless DC motors which provided enough torque for its projected mission. Its primary use was as a "pack animal" but it could be used to carry one astronaut if required. The entire vehicle could be folded up flat.

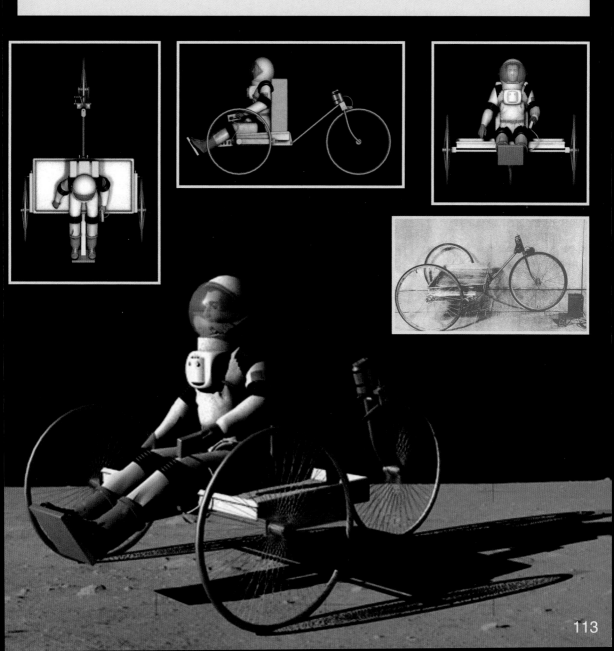

[NAR SLA Mini Base 1969]

Soon after the landing of Apollo 11 North American Rockwell put forward their plans for a small lunar base built using Grumman LM components and North American's own Apollo Command/Service Module. The Saturn Lunar Adapter (SLA) mini base was a clever reconfiguration of the Saturn V SLA compartment to make a large upper stage moon base habitat. Instead of jettisoning panels to reveal the LM ascent stage, the panels would be replaced by an identically shaped habitat module mounted onto a LM descent stage. The first proposal for this appeared in late 1969 and continued to evolve as a potential cheap solution to a moon base until as late as early 1971.

The SLA mini-base (SLAMB) was also adapted to be converted into an orbiting habitat as well as a lunar landing version. Some very curious and imaginative redesigns included placing a modified Service Module upside down inside an appropriate launch fairing and connecting it to the landing version of the SLAMB. In this configuration it would land robotically (without the SM) while a second Saturn V would launch an Apollo CSM atop both a regular LM and an orbital version of the SLAMB. The entire unmanned SLAMB weighed 9315 lbs of which about 1000 lbs was cryogenic liquids. The crew would be launched in an extended LM and CSM, or "quiescent" hardware which could be left in orbit or on the surface while the crew lived in the SLAMB for up to 85 days. This flight profile required a modified LM that could carry all three crew to and from the lunar surface. (see page 121)

Another version called for the SLAMB to be visited by a 2-man LM Taxi while a 24 day or 16 day CSM remained in orbit. All of the SLAMB landers carried a Lunar Rover while at least one carried a lunar escape system, presumably along the lines of the F series Bell escape systems.

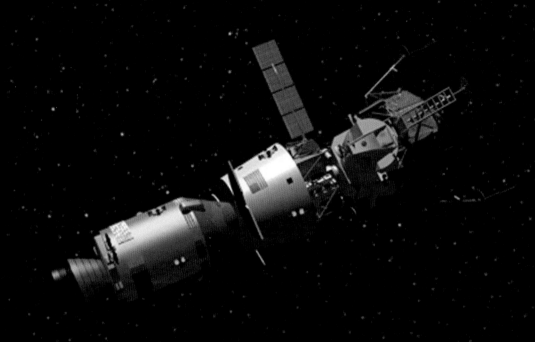

The orbital version of the SLAMB is shown here docked between the Apollo CSM and LM. An air lock/docking tunnel passed through the centre to allow access directly from the CM to the LM.

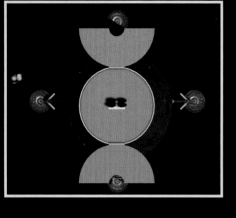

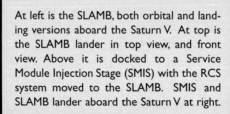

At left is the SLAMB, both orbital and landing versions aboard the Saturn V. At top is the SLAMB lander in top view, and front view. Above it is docked to a Service Module Injection Stage (SMIS) with the RCS system moved to the SLAMB. SMIS and SLAMB lander aboard the Saturn V at right.

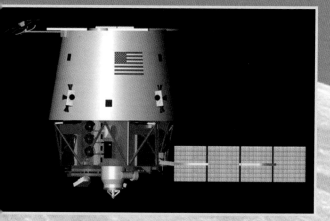
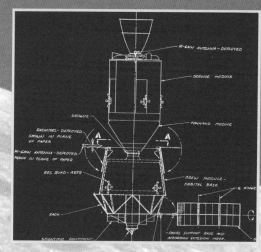
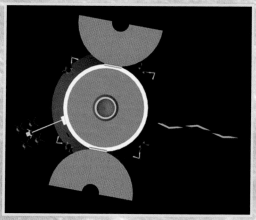
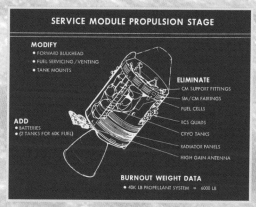

The orbital SLAMB is seen here from three angles plus a diagram showing it docked to the CSM is at top right. The diagram at bottom right shows the dual unmanned launch configuration of the orbital and landing versions. Total payload for trans lunar injection in this version was 99,450 lbs. The landing base weighed just over 9000 lbs with an additional 2780 in consumables and 6800 in cargo, while the orbital base weighed 8970 plus 5330 in consumables and 2000 lbs of cargo. The LM descent stage added a further 22,250 to the total weight. Both the LM descent stage and the so-called "quiescent" SM needed modification for use with the SLAMB.

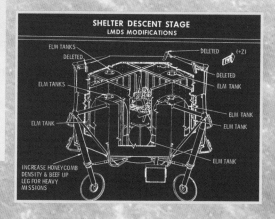

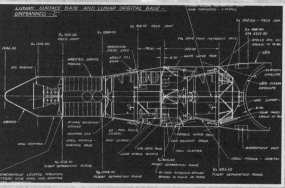

[NAR SLAMB & cryo stage 1969]

Further work continued in 1969 by North American Rockwell to take advantage of the INT-21 version of the Saturn V (used for Skylab) as well as the anticipated nuclear LV-7 third stage. An entire array of permutations of the SLA mini-base were proposed including this version which would be delivered to the lunar surface using a cryogenic landing stage capable of bringing 52,000 lbs of payload down to the surface. The cryo stage had a dry weight of 10,600 lbs and would land robotically while the crew would again arrive using a quiescent 3-man CSM and a specially modified 3-man LM. Surface stay time with this design would be all three crew for 118 days.

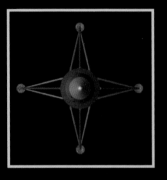

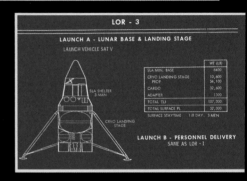
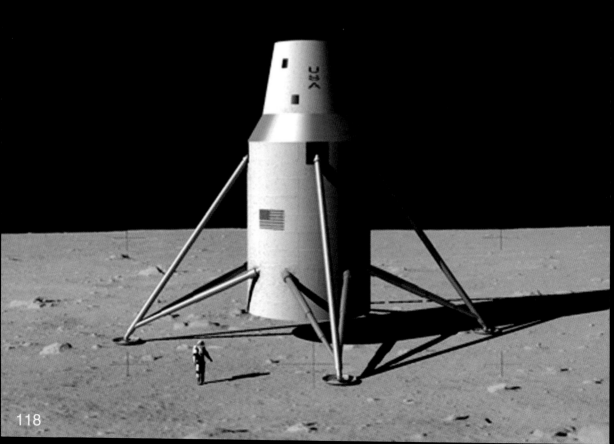

[NAR ADAM 1969]

Another vehicle proposed as part of the SLAMB lunar architecture was known as ADAM. Details on this proposal are so sketchy that even the acronym is a mystery and may stand for Apollo Deployable Advanced Module or perhaps Apollo Direct Advanced Mission. (Not to be confused with the US Army's man-in-space program Project Adam). Regardless, it seems apparent from the diagrams that ADAM was a resurrection of the Lunar Direct proposals of almost a decade earlier. The ADAM vehicle had a modified Apollo command module with a large transparent bubble in place of the docking adapter. The pilot would presumably fly the landing from a standing position with an unprecedented 360 degree view of his surroundings. The CM was mounted atop the LTS (Lunar Takeoff Stage) which weighed 23,300 lbs dry and carried 20,000 lbs of propellant. The LTS in turn was mounted atop the LLS (Lunar Landing Stage) which had a dry weight of 24,200 and 90,900 lbs of propellant. Total TLI mass was 176,000 lbs launched by the INT-21 variant of the Saturn V. A lighter version used the proposed nuclear third stage on a Saturn V and could deliver 160,000 lbs to TLI. The ADAM would be accompanied to the moon by an SLAMB unmanned launch attached to an SMIS (service module injection stage) like the one seen in the centre of page 116.

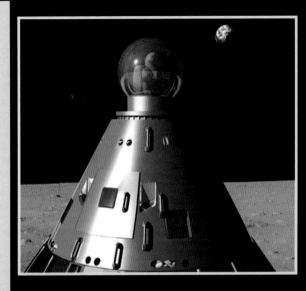

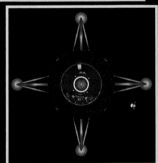

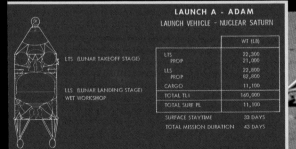

LAUNCH A - ADAM LAUNCH VEHICLE - NUCLEAR SATURN		
		WT (LB)
LTS (LUNAR TAKEOFF STAGE)	LTS	22,300
	PROP	21,000
LLS (LUNAR LANDING STAGE) WET WORKSHOP	LLS	22,800
	PROP	82,800
	CARGO	11,100
	TOTAL TLI	160,000
	TOTAL SURF PL	11,100
	SURFACE STAYTIME	33 DAYS
	TOTAL MISSION DURATION	43 DAYS

[NAR ADAM/SM 1969]

Included in the *Lunar Exploration Planning Studies* of late 1969 it appears that some of the very early designs for Lunar Direct were given a new airing. The North American Rockwell ADAM/SM looked almost exactly like similar proposals from 1962. The advanced landing stage (ADAM) would have a complete Apollo CSM placed atop and would again be launched on an INT-21 Saturn V in tandem with an SLAMB/SMIS launch. In this instance the total TLI mass was 176,000 lbs and 12,300 lbs of cargo could be delivered to the lunar surface. Total mission duration was to be 160 days again using the 3-man LM and quiescent CSM left in orbit. Below can be seen an early model of a very similar design from 1961 along with an accompanying diagram. Wernher von Braun and NASA engineer Gordon Woodcock wrote a paper in 1966 in which they discussed a six-man Apollo system for a *Lunar Direct* mission. This enlarged vehicle would have required a radically uprated Saturn V that would have had gigantic 260 inch solid motors (the AJ260 built by Aerojet) strapped to the S-IC. Unfortunately this 260' diameter motor was too big to transport and so it might have been transported in pieces and then assembled at the Cape or alternatively constructed on or near the launch pad.

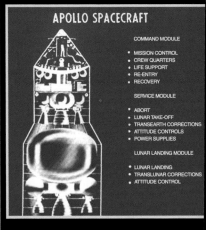

6-man Apollo, cited by Von Braun and Woodcock in late 1966.

[Grumman 3-man LM 1969]

The design for a modified Lunar Module capable of carrying three men to the lunar surface presumably originated at Grumman, although adaptations of this concept were eagerly introduced to plans by both North American Rockwell and Lockheed. The only way such a thing could be accomplished would be in tandem with the quiescent or dormant CSM which would remain in orbit while the crew stayed on the lunar surface. Surviving diagrams of the modified three-man LM are unusually scarce.

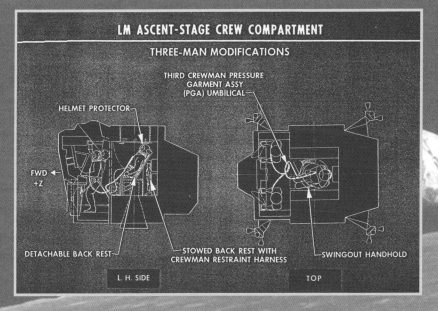

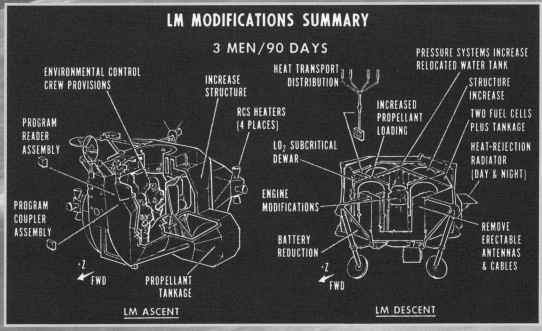

[NAR 6-Man Apollo 1969]

Two Saturn V launches would place two extended lunar modules into lunar orbit using an SMIS boost stage as pictured below left. A specially equipped 66 day CSM would then be launched with the crew aboard a standard Saturn V. Rendezvous in lunar orbit would transfer the crews to the two landers. Variations of this mission design included sending LPM (lunar payload modules) instead of the regular ELM (extended LM) for resupply. The entire S-IVB would remain in lunar orbit attached to the CSM and was designated LASSO (Lunar Application Spent Stage Orbital). The dual ELM/SMIS payload would be launched on a Full J2-S Saturn V with the lighter more powerful J2S engines on both the S-II and the SIVB. The LASSO had a mass of 54,200 lbs, the dormant CSM 22,000 lbs with 12,000 lbs of propellant. Cargo on the LASSO/CSM combination was slated at 15,000 lbs. The dual ELMs had a mass of 70,302 lbs while the SMIS weighed in at 6,700 lbs with another 32,000 lbs of propellant. Cargo was rated at 3,000 lbs. Total TLI mass was 116,000 lbs for the dual ELM/SMIS configuration and 107,000 lbs for the LASSO/CSM.

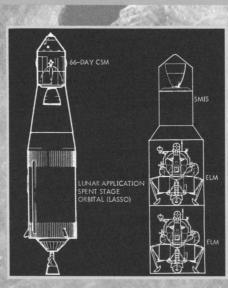

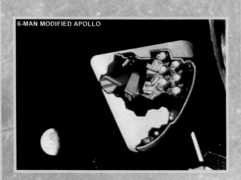

6-MAN MODIFIED APOLLO

[Lockheed 6-Man Apollo 1968]

The Lockheed version of a six-man Apollo landing program looked very much like the North American ADAM/SM configuration. Lockheed also proposed that two new stages be attached to an enlarged Apollo command module but at least one of these new stages in the Lockheed proposal would be powered by a mixture of 82.5% liquid fluorine mixed with 17.5% liquid oxygen as oxidizer and methane as fuel. This extremely powerful and hazardous engine was not yet available when Lockheed made this proposal. The early lunar cargo stage would have a single engine generating 50,000 lbs of thrust and it would have required two of them to effect the landing on the moon. Another version would have used two RL-10 engines powered by LOX and LH2. Perhaps even more hazardous was the proposed earth return stage which would have been driven by fluorine and hydrogen with an estimated Isp of 462 seconds and thrust of 30,000 pounds. The service module was also considered for changes including either a LOX LH2 engine or a FLOX/Methane engine. Lockheed management believed that this advanced architecture could be flying by 1975. The final proposed extra stages were budgeted at $19 million. The chosen landing stage is at bottom right in red.

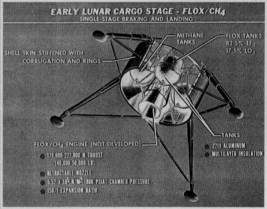

The Lockheed proposal for a 6-man Apollo system looked much like early Apollo Direct systems using fairings to protect the landing legs during launch.

Various different configurations were considered for the landing and Earth Return stages using Fluorine and Methane. The chosen landing stage is below.

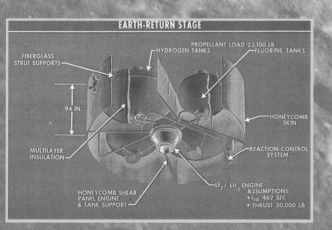

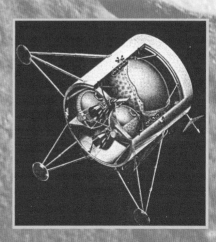

[MIMOSA 1966-1967]

The Study of Mission Modes and Systems Analysis for Lunar Exploration (MIMOSA) was conducted out of MSFC in Huntsville, combining the talents of Lockheed, AiResearch, Bell Aerospace, Bendix and General Electric. The study ran from at least June of 1966 to the final report in April 1967. The resulting in-depth study produced a massive amount of potential uses for the LM and the LLV as well as a host of rover concepts. No less than 23 different lunar habitats were considered along with at least nine different unmanned LM power plants. The habitats began as small as the Goodyear *STEM* inflatable and proceeded through the *LM Shelter* all the way up through an enormous 12-man 18,143 kg habitat (see 2327-01 opposite). At least ten of the habitats were derivatives of the remaining thirteen or were simply off-loaded versions, removed from their landing stage. Appearing here are the remaining proposals that have not appeared elsewhere in this book along with the nine LM Truck Power Plants.

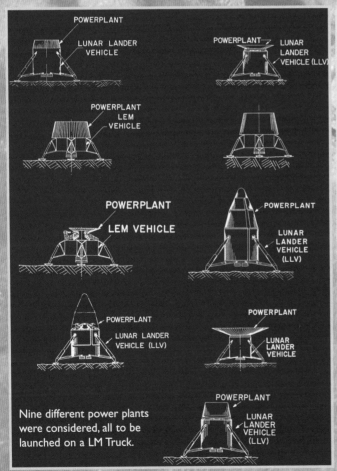

Nine different power plants were considered, all to be launched on a LM Truck.

MIMOSA Lunar Habitat configurations (see opposite):

2321-01 14 day LM Shelter 2-man
2321-02 30 day ELS 2-man
2321-03 8 day STEM 2-man
2322-01 30 day Growth ELS 3-man
2322-02 90 day 4.4m Module on LMTU 3-man
2322-04 180 day Extended ELS 3-man
2322-07 90 day 4.4m Module 3-man
2325-01 180 day LESA 6-man
2325-03 180 day Nuclear LESA 6-man
2325-05 180 day 2 stacked 4.4m Modules 6-man
2327-01 180 day 10m Module 12-man
2327-04 180 day 2 stacked Nuclear LESA 12-man

All of the above were to be launched on either a LM, an LLV or a LM Truck with the exception of 2322-02 which would have required something called a LM Truck U, presumably for uprated. This enlarged LM Truck appears to have been too large to fit in the standard Saturn V SLA since the module designed to sit on it was 4.4m in diameter and the descent stage seems to be larger than that.

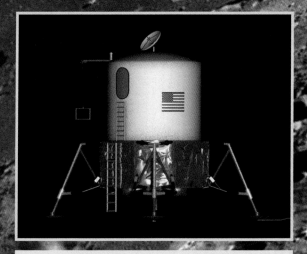
2322-02 4.4m Module on LM Truck-U

2325-05 2 stacked 4.4m Modules (nuclear)

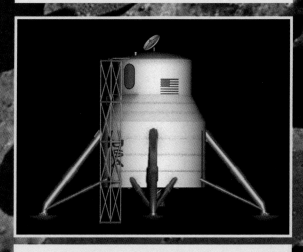
2322-07 4.4m Module on LLV

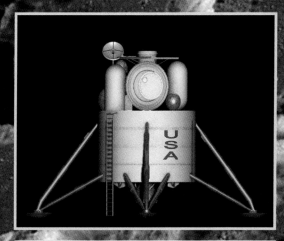
2322-04 Extended ELS

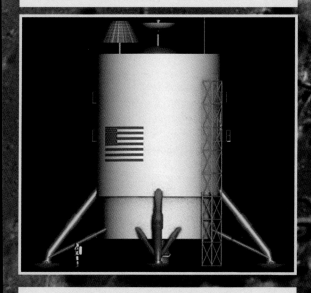
2327-04 2 stacked LESA (nuclear)

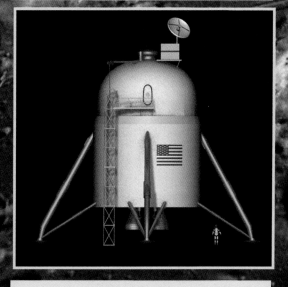
2327-01 10m Module

[MOBEV 1966]

LUNAR SURFACE MOBILITY SYSTEMS COMPARISON AND EVOLUTION (MOBEV)

FINAL REPORT

VOLUME II
BOOK 7
EVOLUTION METHODOLOGY
USER'S MANUAL

BSR 1428

NOVEMBER 1966

 Aerospace Systems Division

Rocket flying devices were a particularly popular idea in the 1960's and they would, along with lunar rovers, come under considerable scrutiny. The most thorough report conducted on both of these methods of lunar transportation was undertaken jointly by *Bendix*, *Bell* and *Lockheed* and revealed in November 1966 under the title "Lunar Surface Mobility Systems Comparison and Evolution" or MOBEV. It came hard on the heels of Bendix engineer Richard E. Wong's August 1966 paper entitled *Lunar Surface Mobility Systems*. This multi-volume study included everything that was under serious consideration at the time. Beginning with the simple adaptation of existing pressure vessels.

To some it was believed that the cheapest and simplest way to create lunar mobility was to put Apollo on wheels. The MOCOM would be a Mobile Command Module and the MOLEM would be a Mobile LM ascent stage. Since NASA had already approved the pressure vessel designs for the Apollo modules it seemed to make perfect sense to place them aboard a suitable chassis with wheels.

This simplest design was given the acronym of LSSM for *Local* (sometimes *Lunar*) *Scientific Survey Module*. Over the years this acronym would be mistakenly applied to many different shapes and sizes created by several different contractors, primarily the teams of *General Motors/Boeing* and *Bendix*.

The primary *Bendix* chassis would have four wheels and would sport a variety of different wheel spans. This would allow it to be adapted to a host of different configurations ranging from extremely small unmanned exploration rovers, to trailers, to large multi-crewed pressurized vehicles for long-range exploration. The *Boeing/GM* chassis design had a unique flexible frame and was to have been used for everything from simple one-man exploration rovers to a large MOLAB (mobile laboratory). Since none of these designs ever flew, nothing was ever finalized regarding the preferred method of deployment for all of these vehicles. However, there does seem to have been some consistency for a deployment architecture between *Bendix* and *Grumman*.

Since as early as 1959 serious thought had been given to a suitable means of transportation on the lunar surface. The early rocket pioneer, Hermann Oberth had made proposals for a "Moon Car" which involved a highly unlikely vehicle standing on a single leg. The design was quite peculiar and had absolutely no chance of ever being seriously considered for flight.

In the early 1960's many different designs came forth from *Bendix, Boeing, General Electric, Grumman, Hayes* and others. The first order of business was to settle on a wheel that could work on the moon. Rubber tires, inflated or not, were out of the question, they couldn't easily handle the extremes of heat or pressure. Therefore it was decided that some sort of metal wheel would be the best solution. *Grumman* seemed to favor a spiral spring "resilient" design for their early large pressurized MOLAB, but they also used a dome shape with radial teeth for at least two of their Earth-bound LRVs. Both were conceived by *Grumman* engineer Edward Markow and had come out of extensive research to try and create a wheel with highly flexible characteristics. They were said to have "the properties of large footprints for weak soil, low unsprung weight to accommodate the dynamics of reduced lunar gravity, and invulnerability to micrometeorites and low temperatures." Ultimately the "resilient wheel", which Markow had adapted from a 1942 design, was used on the Mars Rovers over three decades later.

Bendix favored an array of metal circles for their wheels, designed to take the weight but compress like springs as they passed over the rough lunar surface. The *Bendix* "coiled spring" rover and the *Grumman* "elastic-conoid" LRV would compete directly against one another at a USGS site in Arizona in September 1969. Neither would end up being used.

Boeing and Hayes both seem to have hit upon a wire wheel as a suitable solution. It didn't require inflation, it weighed very little and, given the right design, could take the load of a fairly large mobile laboratory with a crew of at least two. Some of the *Boeing/General Motors* team benefited from working with *Goodyear* whose experience with wheels was obviously considerable. It was apparently *Goodyear* who first came up with the idea of weaving a sturdy gauge of wire together to create the "tire" part of the wheel. This version of a resilient wheel was patented by John Calandro and Norman James for NASA. *General Motors* engineer Bob Petersen suggested the inclusion of a herring-bone pattern of interlocking steel strips to improve the performance. Many of these designs would be tested here on Earth, long before they had any chance of being sent to the Moon or Mars.

Different wheels for the moon. "Resilient" wire wheel, Bendix is 2nd from left and Markow's "elastic conoid" is at far right.

[Bendix & GM SLRV 1967]

The testing site used by the *US Geological Survey* was a full-size replica of the proposed primary Apollo landing site. This is a remarkable story in itself. It seems that between July 28th-31st 1967 the USGS installed a considerable amount of ammonium nitrate fertilizer around a stretch of barren lava in Arizona known as *Cinder Lake*. The team detonated these fertilizer explosives to create an extremely accurate replica of the moon. They then returned in October and did this again, increasing the size of the crater field. This astonishing man-made formation can still be visited, it is not far from Flagstaff Arizona, but for those with less air-miles, try checking it out on Google-Earth at 35°19'18.75" N by 111°31'01.91" W.

Once the artificial crater field had been completed, the USGS team then proceeded to train the astronauts in both geology and in the foibles of driving the bizarre array of potential vehicles provided by *Grumman*, *Bendix*, *Boeing* and others.

The *unmanned* prototypes tested by the USGS included competitive designs from *GM* and *Bendix*. The *GM* version used wheels (at this point they appear to have been rubber tires although a mock-up seems to have been nothing more than cloth stretched over wire) while the *Bendix* version used tank-tracks. *Bendix* had considered three different power sources for their SLRV *Surveyor*, a small SNAP-11 RTG and two different configurations of solar cell arrays. As early as May of 1964 the two competing designs were put through their paces on the Bonita Lava flow in Arizona. Unfortunately for *Bendix* their tracked design didn't hold up and *GM* won the day. This competition eventually produced results that showed that these two *Surveyor Lunar Roving Vehicles* (SLRV) were inadequate for their defined task of landing on the moon and then rolling around looking for suitable landing sites for subsequent manned landers.

A very early model of this GM design appeared in Life Magazine, April 27 1962, with large yellow wheels.

The Bendix SLRV Surveyor with tracks is seen opposite and in the drawing at right.

The GM prototype with cloth wheels is at top and the competition version with regular wheels is above.

Both of these were briefly considered for the advanced Surveyor program but neither would fly.

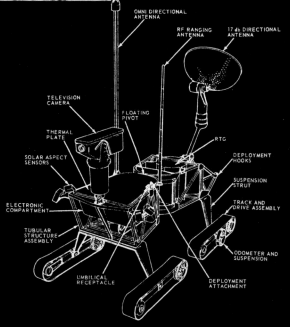

129

[USGS LMs 1964-67]

Another of the interesting off-shoots of this particular training program was the deployment of what must surely be the strangest pair of Lunar Modules ever built. One was commissioned specifically by NASA for the USGS team, and built to scale by *Grumman* in the summer of 1964. It was made entirely of plywood and its main mode of propulsion was…flat-bed truck. This was used by Dave Scott to plan the SEVA on Apollo 15. The other, which arrived several years later, and was much easier to move around, was little more than a LM shaped tent and was deployed at the top of a ladder at *Cinder Lake*. Both were designed purely to give the astronauts a basic feel for working with a home base, while using a familiar shape. Neither LM ever left the ground…

The USGS cloth LM with Gene Cernan and Jack Schmitt (right) and with John Young and Charles Duke (top right)

The USGS plywood LM is delivered by truck (above centre left and right)

[Bendix Specified LSSM 1966]

The basic *Bendix* chassis used for many of the rover studies was a simple flat platform with a driving seat up front and four large coiled wheels. This was used in simple form as their Specified LSSM. It was supposed that this chassis and wheel assembly could very likely be folded up and placed on board the *LM Truck*. Several NASA photographs incorrectly identify this as having been built by *Grumman*. The dimensions of the Specified LSSM were identical to the R1BE LSSM (page 171) with the exception of the wheels which were 2 inches smaller in diameter. Other simplifications included less control displays, simpler steering mechanisms, and no RTG. The Specified LSSM was to have been part of the Apollo Applications Program and would have weighed 900 pounds,

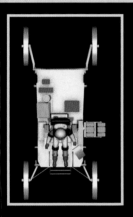

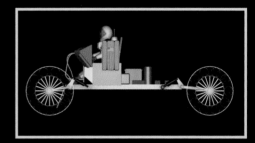

The Bendix Specified LSSM mock-up is seen at right and in diagram form (below).

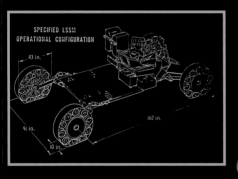

The Bendix LSSM is seen here with an LLV in the background

[Bendix MTA BX-1 1966]

The Bendix Mobility Test Article BX-1, with its giant 80" titanium alloy wheels, was tested at Yuma Proving Ground between September 1966 and February 1967. The tests were done to obtain quasi-steady state mobility performance data under simulated lunar conditions. This extremely large chassis would have been used for the later Bendix MOLAB designs. It was shown to be capable on slopes of up to 60 degrees over a six mile course. Wernher von Braun inspected the MOLAB chassis (above) during some tests in Huntsville.

[GM MTA GM-1 1967]

The *Boeing* LSSM, with its flexible frame, would evolve from the *General Motors* tests performed in September of 1966. It was effectively a scaled-up version of the SLRV that had been tested out in Arizona two years earlier. GM had won the competition with *Bendix* and so now a larger version would be put through its paces on the same proving ground where the world's first *Jeep*, the *Bantam LRV*, had been tested exactly 26 years earlier—the US Army's Aberdeen Proving Ground in Maryland. The original GM SLRV had always had what appeared to be a split chassis and six wheels. This was an illusion since it never really had a chassis in the conventional sense. The beauty of this design lay in the fact that the wheels and axles were attached to a frame that could twist. This provided unprecedented mobility over rough terrain. The test vehicle which showed up at Aberdeen in the fall of 1966 had evolved to the point where the rubber tires had been replaced by wire, but Petersen's herring-boned straps were not yet in place. This vehicle, known as the 6 x 6 Mobility Test Article (MTA GM-1) had six metal treads that wrapped around the circular perimeter of each wheel. The MTA with only the basic frame needed to house a driver and motors, weighed only one-sixth what the whole vehicle would have weighed, had a MOLAB pressure vessel been placed aboard. It was therefore considered a good simulation of how the whole package would have performed on the Moon. Should the MTA prove effective it would have been adapted for the GM/*Boeing* MOLAB, while a smaller scaled-down version would have been used for the GM/*Boeing* LSSM. Because the flexible frame looked like a four-wheel vehicle pulling a two-wheel trailer it was sometimes referred to as a Dual-LRV. *Bendix* also came forward with a dual-LRV that looked very similar to the *Boeing*/*GM* version but was to have used the *Bendix* wheels.

The GM Mobility Test Article arrives at Aberdeen Proving Grounds where the first ever Jeep was tested. Note rubber inner tubes at top and plain wire wheels below. These inner tubes were added to minimize wheel damage from excessive individual wheel loads. Also note the flexible frame. The GM-1 weighed 1680 pounds which was about one fifth of the expected mass of the operational MOLAB.

[Boeing LSSM 1966]

The final version of the Boeing/GMDRL LSSM was submitted to NASA in June 1966. Boeing took their flexible frame chassis and loaded it up with scientific equipment. The full six-wheel LSSM would have carried a gravimeter, ground truth package, drill, penetrometer, theodolite and ranging laser, electromagnetic coils and boom, magnetometer and a nuclear experiment. The LSSM would fly as part of the Apollo Applications Program on board a LM-Shelter. It was capable of sorties of up to six hours over an 8 kilometer radius of the landing site. It could carry additional PLSS backpacks and a second crewman if necessary. The rear trailer would carry communications, navigation, drive system electronics and power. Round trips of up to 26 kilometers were possible. Each wheel was individually driven by an electric motor. What is also of interest was Boeing's design for an unloading concept. This would have required the LM-Shelter to have been equipped with a top-mounted crane, placed directly over where the original LM docking hatch would have been.

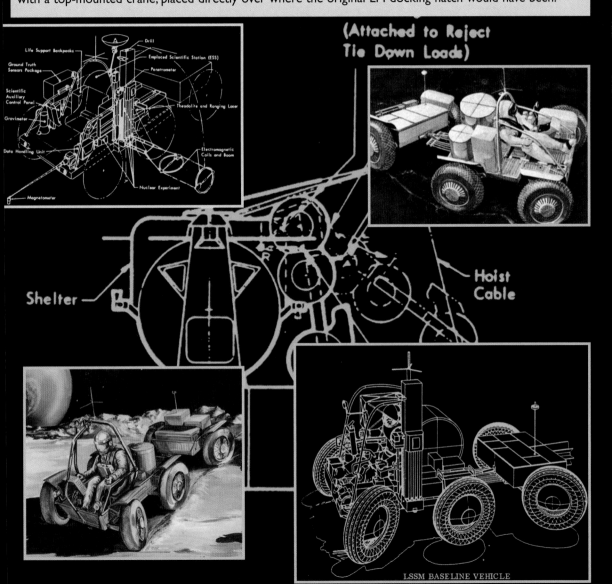

[Boeing Specified LSSM 1966]

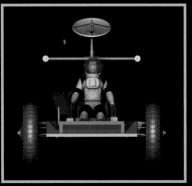

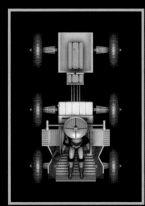
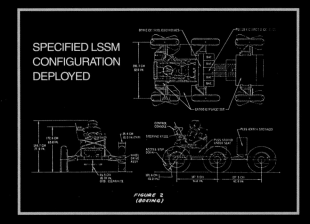

In November of 1966 Boeing would submit their design for the lower weight *Specified* LSSM. This six wheel flexible frame concept would have a 120" wheel base and 82" width. It was capable of carrying one crew with life support and a 700 pound equipment package. The 40 inch wheels were made of wire, similar to that which ultimately flew on Apollo 15.

[Grumman Dual LRV 1969]

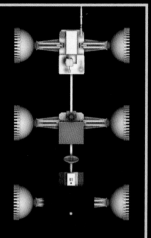

The concept of independent frames for each axle seems to have also been adopted by *Grumman* who in turn built a mock-up of their own six wheel vehicle, they called this a "land-train" but this also became known as a Dual-LRV. This large *Grumman* rover was also depicted with the spring-wheel concept in the artist's rendering seen on the next page. The six-wheel *Grumman* vehicle featured a giant cooling radiator panel that acted as a sunshade over the driver's seat. This vehicle was capable of driving in either direction and was able to operate in an unmanned mode. It would have been dispatched to the moon aboard an Extended LM. Creative use of packing space would have allowed the entire vehicle to have been folded and stored inside either one or both of the front quadrants of the LM descent stage (below).

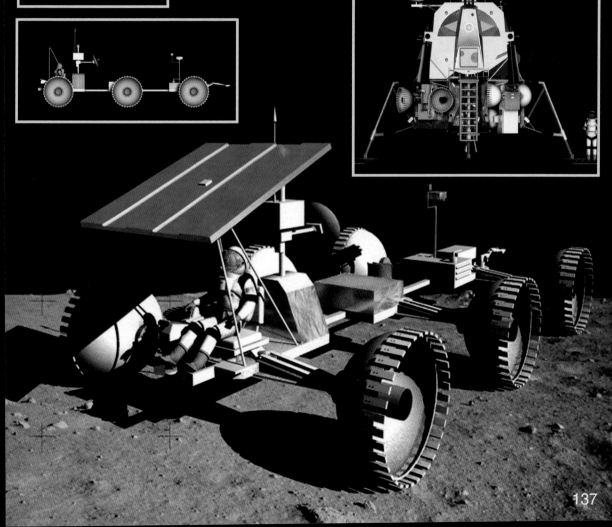

In October 1969 the Grumman Dual LRV was tested against the Bendix version for use in missions beyond Apollo 20. It was found to be 50 pounds overweight and may have had problems mastering slopes over 35°. It had a 96" wheel base and was 230" long and weighed 700 pounds. It was powered by a SNAP 19 reactor.

Note different Grumman wheel designs above and at right.

The full size Grumman dual LRV mock-up (below)

[Grumman LRV 1969]

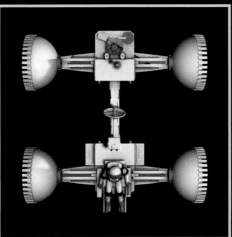

Grumman's gigantic six-wheeled roving vehicle was a natural offshoot of their regular four-wheel rover. Both were built in full-size and the four wheel version was used in the 1969 contest against *Bendix* out at Cinder Lake. The full-scale mock-up did not have the SNAP 19 reactor (below) that would have powered the lunar version. The Grumman LRV could be modified to carry two men (see diagram).

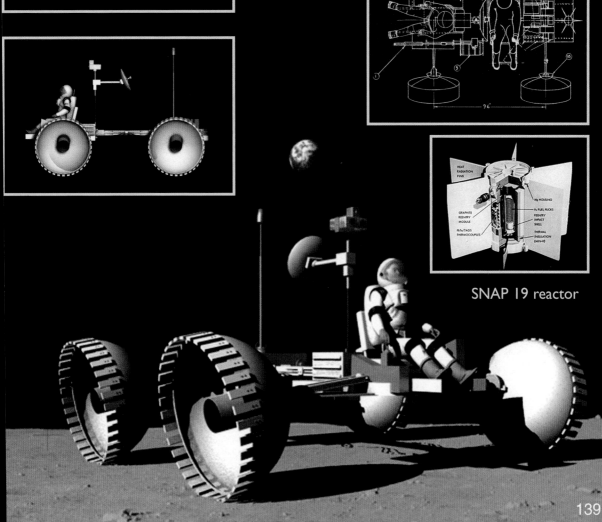

SNAP 19 reactor

The Grumman LRV is tested by an assortment of drivers including astronaut Harrison Schmitt (centre left) at Cinder Lake

[Grumman LRV 1964]

Grumman also had ideas for a much smaller rover which would have been flown to the Moon strapped in two parts to the back quads of the LM. The two parts would be removed by the crew and then combined to make one vehicle. This concept seems to have been considered in two possible configurations, the first one would have a sun-shade, similar to that on the large Dual-LRV while the second had a rather unlikely inflatable cockpit. The sun-shaded version was depicted as the rover of choice for *Grumman's LM Shelter* concept.

[Grumman Inflatable LRV 1964]

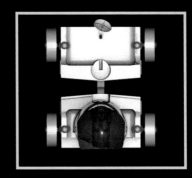

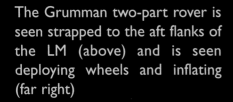

The Grumman two-part rover is seen strapped to the aft flanks of the LM (above) and is seen deploying wheels and inflating (far right)

Some artistic license has been taken here to show the inflatable bubble with a gold reflective finish. The author has not been able to determine exactly how the surface of the bubble would have appeared.

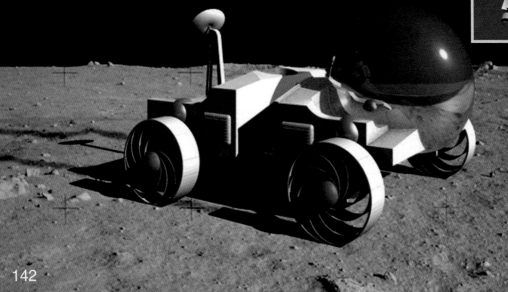

[Bendix Dual LRV 1970]

In January of 1970 Bendix filed their final design for a Dual LRV with NASA. It was a six wheeled configuration which would be stowed in two LM descent stage quadrants (right) and coupled together by the astronaut during the unloading operation. This lightweight design could carry two men along with the Lunar Surface Drill, a rock box, the Lunar Hand Tool carrier, a gravimeter, survey staff, TV camera, facsimile camera and magnetometer. The vehicle could be powered by a small RTG carried on the rear trailer. It was 94 inches wide and 58 inches from wheel to wheel. This vehicle evolved from Bendix's 1968 study to provide a small vehicle capable of being launched as part of the Extended Apollo program. The original specifications would have provided for an RTG powered unmanned version.

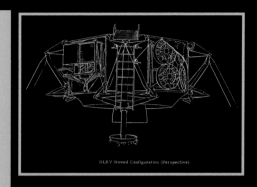
DLRV Stowed Configuration (Perspective)

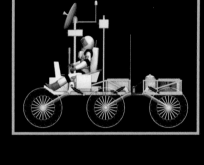

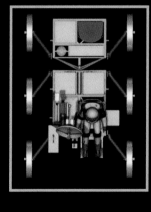

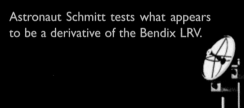
Astronaut Schmitt tests what appears to be a derivative of the Bendix LRV.

143

[NAA Winged Apollo 1967]

By 1967 the idea of a reusable spacecraft had already caught the attention of the US Congress. *North American Rockwell*, contractor for the Apollo CSM, decided that it would be possible to modify the entire spacecraft into a reusable "space shuttle". With the addition of two protruding fairings and an advanced ablative coating (based on the X-15A2) it was thought that the Apollo CSM could be retooled to fly like a lifting body. As extraordinary as this might sound, there is a certain elegance to the idea. It would have fold-out titanium wings inside the fairings so that it could continue to fly inside the atmosphere after the period of heating was over. The nose of the CM would be rounded for better heat dispersion. Even more ingenious was the notion of a discardable main engine bell so that the rocket could continue to provide thrust for a powered descent and landing by reducing the expansion ratio. The reaction control thrusters and landing skids rotated out from behind the service module fuselage and a retractable nose gear provided three point landing capability. The service module would have been retooled to include two large hatches that could open to reveal a mini shuttle-type cargo bay.

[TRW LLV/LLS 1963]

At least one of the lunar surface vehicles was so large that there was simply no way that it could be sent to the moon aboard a *LM Truck* and so it would have required a much larger lander called an LLV or *Lunar Logistics Vehicle*. The LLV was also the prime candidate for transporting the proposed large moon base habitat modules known as LESA. Thompson Ramo Wooldridge Corporation undertook an intense study at the end of 1963 and came up with eight permutations of the LLV including four manned versions. It was their intention to use a combination of descent stages (called L-I, L-II and L-III) to allow for a variety of possible LLV landers. The four manned versions would have had an Apollo CM stacked on top.

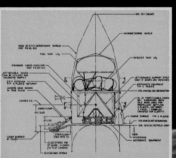

The unmanned LLVs shown in these renderings are equipped with habitation modules. They could also carry large MOLAB or construction vehicles.

The eight TRW LLVs included the SB-625 (top left) a two-stage configuration utilizing two RL-10 engines in the L-I stage and three throttleable LM engines in the L-II stage. The two-stage SB-713 (centre above) a two-stage vehicle utilizing two RL-10 engines in the L-I stage and a single RL-10 engine in the L-II or return stage. The SB-910 (top right) a modular manned three-stage configuration that utilized a module in each stage, however, the entire L-I and L-II stages were identical. The SB-908 a hybrid three stage manned configuration with two RL-10s on L-I which was jettisoned in lunar orbit. L-II was almost identical to L-I on this configuration but has the landing legs attached. Next was a tailored L-II with a Lunar Landing Module connecting it to the second L-II but this LLM had no legs. Finally an RL-10 was used for ascent. Once on the moon it looked exactly like the SB-910. The SB-809 a modular LLS configuration (above left) powered by two RL-10 engines with two stages, L-I and L-II. The SB-529 (diagram left) a "Tailored" LLS single stage low cargo configuration that used three RL-10 engines for landing. The SB-810 utilized a tailored L-II stage and a modular L-I stage. Once on the moon it would have been indistinguishable from the SB-625. And finally the SB-911 (left) which utilized a tailored L-II stage from the SB-713 and a modular L-I stage. The tailored L-II stage in this case included a Lunar Landing Module which was left on the lunar surface. Two RL-10 engines were used for the L-I stage but it was jettisoned during descent. A single RL-10 was used on final descent and again for ascent.

[Douglas/IBM LASS 1966]

A proposed larger version of the LLV was called the *LASS (Lunar Applications of a Spent S-IVB/IU Stage)* this version was literally an adapted Saturn V third stage with legs added. In its primary iteration the LASS would be powered by a single J-2 engine. The LASS payload adapter was over 28 feet high. This and the LLV concept seem to have originated out of a study contract awarded to STL, Grumman and Northrop in October of 1962. Nineteen different companies had competed for the contract. The original parameters were for a 9,000 pound vehicle launched on a Saturn I and capable of carrying 1,500 pounds of payload, and a 90,000 pound vehicle carrying 20,000 pounds of payload, and launched on a Saturn V. These two different specifications would have been capable of carrying entire lunar habitat modules or very large rovers. The LASS was studied in at least six different variants. The one pictured

on the page opposite has the J2 along with two gimbaled RL-10A vernier engines, this version was called configuration 1A and increased the LASS payload from 26,940 lbs to 27,300. Configuration #2 (seen below) inverted the entire vehicle with the legs launched inside a fairing instead of strapped down under thin shrouds against the S-II/SIVB interstage. This would still have used the J-2 to power translunar injection but would then have used a single RL-10A engine to land on the moon. This version had a payload limit of 21,840 lbs. Configuration #3 would have two RL-10A engines mounted at right angles to the J-2 and would allow the entire LASS to land horizontally on skids. Configuration #4 would actually stage the S-IVB just above the black line seen below and would reveal an LLV beneath (similar to that on page 145). This version could deliver 25,340 lbs to the lunar surface.

LASS drawing (above) and LASS observatory (below)

In its various permutations the LASS would have been able to land either horizontally or vertically (using either skids or legs accordingly). In its horizontal version it would have acted like a huge garage with rovers and possibly MOLABs to drive out of the top to explore the lunar surface. This version would have needed a LM companion to act as a shelter until the huge LH2 tank could be converted into a habitable shelter. The cargo in the payload area would provide the necessary provisions to accomplish this task. In its most optimistic version, several LASS would be connected by pressurized tunnels to make a complete moon base. Another version of this horizontal mode would have accommodated a large telescope mounted in the side of the LASS to use as a permanent lunar observatory. Finally, yet one more version would have been designated LASSO, and would have remained in lunar orbit, acting as an orbiting lunar space station.

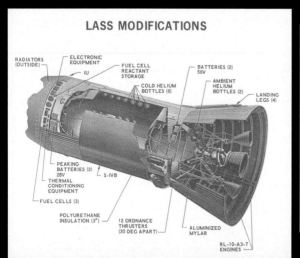

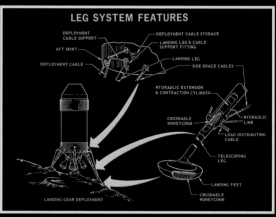

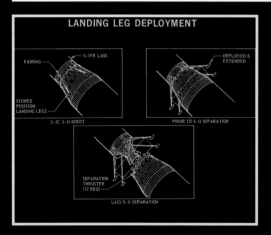

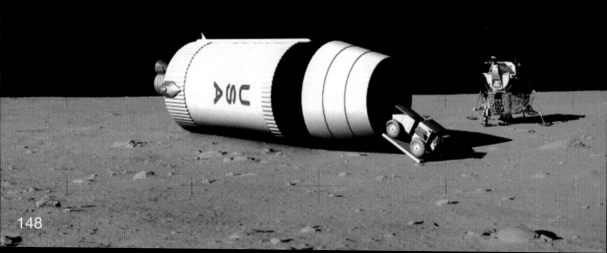

[NAA SMLV 1966]

Another LLV under consideration was the *Service Module Logistic Vehicle* proposed by *North American*. This was basically an Apollo Service Module extended in length by about four feet with a landing gear added. *North American* thought they could put 11,000 pounds of payload on the moon with this concept but it would require a new throttleable engine. The LM descent engine was considered as a possible replacement for the regular SM engine. Other plans called for a modified *Agena* engine.

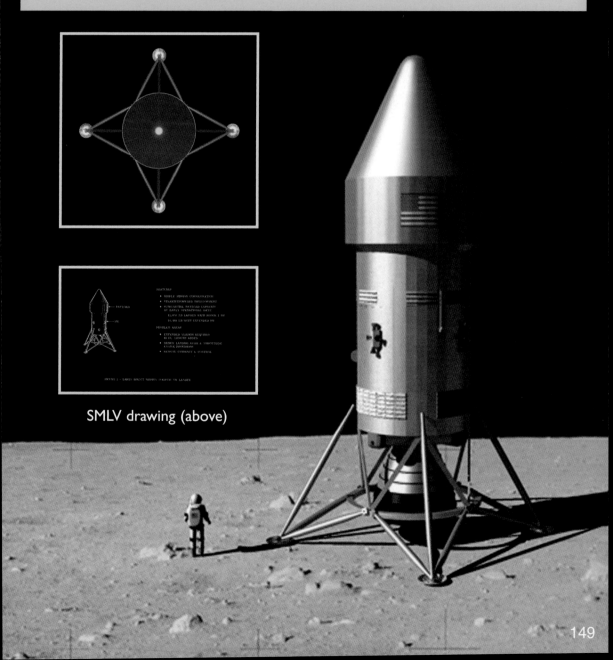

SMLV drawing (above)

[Grumman MOLAB 1965]

Perhaps the most famous of the large pressurized rovers was the *Grumman* MOLAB. This sleek white 6,500 pound capsule was to have been dispatched to the moon aboard a *LM Truck* descent stage. Through an ingenious folding design this would have probably been the largest payload lofted on the *LM Truck*. The *Truck* would arrive, in an unmanned state, where it would be met by astronauts who had arrived via a *LM Taxi* flight. After landing the MOLAB would unfold like a concertina, revealing its train of cargo trailers and full length of 42 feet. All of the *Bendix* and *Grumman* MOLABs would have slid forward on a vertical rail gradually lowered in front of the cabin. The rail created a ramp down which the vehicle could then drive. This folding railed deployment mechanism seems to have originated at *Bendix* and was also their preferred method for deploying their own assortment of roving vehicles. *Grumman* built an Earth-bound MOLAB which was a full size replica of the actual lunar vehicle, with spiral spring wheels and equipped with one of its potential trailers.

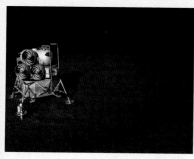  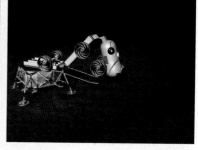

The Grumman MOLAB deploys along its rail mechanism

 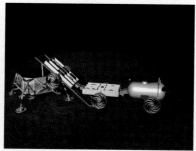 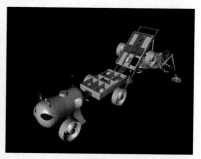

The 31 foot long Grumman Mobile Base Simulator for MOLAB.

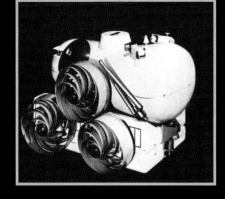

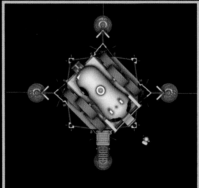

MOLAB shown folded as it would be on the LM Truck (above)

Full-size mock-up awaits completion (above). Contractor model (below)

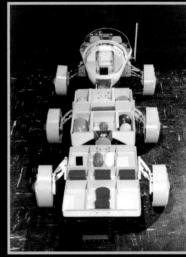

[GM LMDV 1965]

Another MOLAB, with conventional rubber tires and built by *General Motors* at a cost of $600,000, was shipped off to the USGS in Arizona to be used for training at Cinder Lake. This version was variously named the *Mobile Geologic Laboratory* (by the USGS), the *Lunar Mission Development Vehicle* (by GM) and just MOLAB by practically everyone else. It was capable of running around for a week with two men on board and at different times it was used to test a gamma-ray spectrometer, a total field magnetometer, a magnetic susceptibility bridge, a stereo periscope, an aircraft sextant, and a gyrocompass, but ultimately it was considered inherently unstable by many of those who worked with it.

The USGS team with the LMDV

[USGS Explorer 1967]

Two of the vehicles that went through extensive trials by the USGS were the *Explorer* and the *Grover*. These two test-beds were actually constructed by the team at the *USGS Branch of Astrogeology* as part of the Apollo Applications Program (AAP). The *Explorer* was first deployed into the field in the summer of 1967. It was a four-wheel drive vehicle, much larger than anything that would actually fly. In some respects it was similar to the basic fixed chassis vehicle known as the *Bendix* LSSM. It would be tested well into September of 1967 at Cinder Lake and proved capable of easily climbing in and out of the man-made craters. It continued to be used for basic training even in 1972.

The USGS Explorer during training and with John Young (top right)

[USGS Grover 1970]

The *Grover* was the nick-name for the USGS prototype of the final GM/*Boeing* LRV. It was designed to operate in one gravity and was thus called, variously, the 1G Rover and the USGS *Geologic Rover*. It was built closely following the blueprints of the actual Apollo LRV that *Boeing* was preparing for flight, but it was constructed in only three months at a cost of $1,900.

The GROVER takes Dave Scott and Jim Irwin around Cinder Lake

[GM/Boeing 1-G Trainer 1970]

A similar 1-G test bed built by *Boeing*, which apparently had many design flaws, cost over $1.25 million and was almost never used to train the astronauts. Meanwhile, the *Grover* would become a popular favorite of the Apollo astronauts during their training in Arizona.

Note the different wheels. Above the trainer has tires, while below it is equipped with the wire wheels used on the moon.

[Grumman MOLEM 1966]

Returning to the MOBEV studies of 1966 we find that *Bendix* examined the feasibility of putting an existing pressure vessel on wheels. The three choices were MOCOM, MOLEM and a third option based on a *Boeing* habitation module known as the *Multi-Mission Module* or CAN; thus MOCAN.

The MOLEM would be a modified *LM Shelter*, placed on the moon atop a *LM Truck*. *Grumman* had also pitched their own variant of MOLEM which would have used the metal-elastic spring wheels favored by the LM manufacturer. The *Grumman* MOLEM would probably have retained its RCS thruster system because it was cheaper to take that route, this became known as the "minimum-change" configuration. The Bendix version of MOLEM, however, made no such concession and showed a highly modified ascent stage without thrusters placed atop the Bendix LSSM chassis. Presumably this would have required an RCS equipped *LM Truck* as its lander; the same as MOCOM and MOCAN.

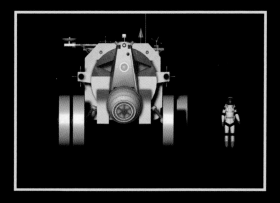

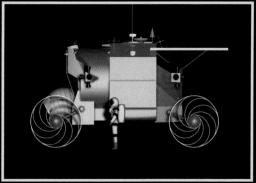

[Bendix MOLEM 1966]

The *Bendix* MOLEM cabin was a modified LM ascent stage and was to be mounted on an aluminum box-beam frame supported by four spiral metal-elastic wheels. The suspension was folded to provide a minimal storage configuration for unmanned delivery on the *LM-Truck*. The environmental control system (ECS) radiators, S-band communications antenna, and a radio direction-finder loop antenna were located on top of the cabin. Scientific equipment lockers, radioisotope thermoelectric generator (RTG), batteries, liquid hydrogen tank, fuel cells, primary power accessories, and the mobility control unit were to be located on the aft platform. Liquid oxygen tanks were located on each side of the cabin.

The MOLEM would have carried a crew of two with sufficient life-support capability to complete a 14-day lunar stay with a 7-day emergency reserve. Accommodations were provided for 320 kg of scientific equipment, and 75 kwh of power was budgeted for the equipment. This vehicle had a total range capability of 400 km with a typical mission radius of operation of 80 km.

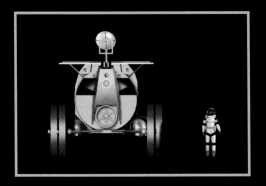

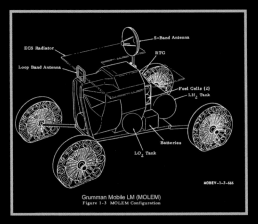

Grumman Mobile LM (MOLEM)
Figure 1-3 MOLEM Configuration

[Bendix MOCOM 1966]

The basic Apollo Command Module was examined by Bendix to determine those subsystems and components required in a mobility system to create the MOCOM concept. The propulsion system, the stabilization and control system, and the re-entry systems were not required in the mobile unit. Another variation considered was the Command Module shelter which would have been a modified Apollo Command Module, similar to that used in MOCOM but placed as a static shelter atop a LM Truck descent stage.

The MOCOM utilized the modified Apollo Command Module cabin which was adapted to mount on a four-wheel, rigid-chassis mobility system (a *Bendix* LSSM). The system was delivered unmanned on the *LM-Truck* with suspension folded. Six months of dormant storage on the lunar surface were provided for, along with remote checkout, unloading, and driving features. The cabin was mounted on the chassis with the major axis of the CM cone in a vertical position. This was done to accommodate the design of the docking adapter, and since the major load paths (structural strength) were in this direction, the astronaut position in the cabin, and internal furnishings and operational equipment, were revised to compensate for the attitude change between the MOCOM and the CM. Two liquid hydrogen tanks were mounted under the radiators. Two spherical liquid oxygen tanks were mounted in the right and left rear corners of the vehicle.

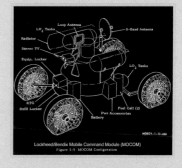

Lockheed/Bendix Mobile Command Module (MOCOM)
Figure 1-5 MOCOM Configuration

[Bendix MOCAN 1966]

The MOCAN used the larger LSSM chassis, designed for the MOLAB, and the *Boeing* CAN basic structure. The *Apollo Multipurpose Mission Module* with the completely impenetrable acronym, CAN, was to be mounted on a rectangular aluminum box-beam frame and supported by four 203-cm (80-in.) diameter metal-elastic wheels. These were the largest of the wheels considered in the tests made around Huntsville. The CAN had been designed to utilize fully, the volume on top of the *LM-Truck* and within the SLA. MOCAN with a free volume almost five times greater than desired, provided relatively unlimited capability (compared to MOLEM and MOCOM) for extensions in mission duration or increases in crew size.

As interesting as these three vehicles might seem, none of them went beyond this *Bendix* study because, *"Each of the derivative concepts studied is technically feasible, in the case of MOCAN with degraded performance, and in the case of MOLEM with marginal cabin space. However, they do not offer any measurable advantages over a system which is designed specifically for mobility use."* And so MOLEM, MOCOM and MOCAN disappeared.

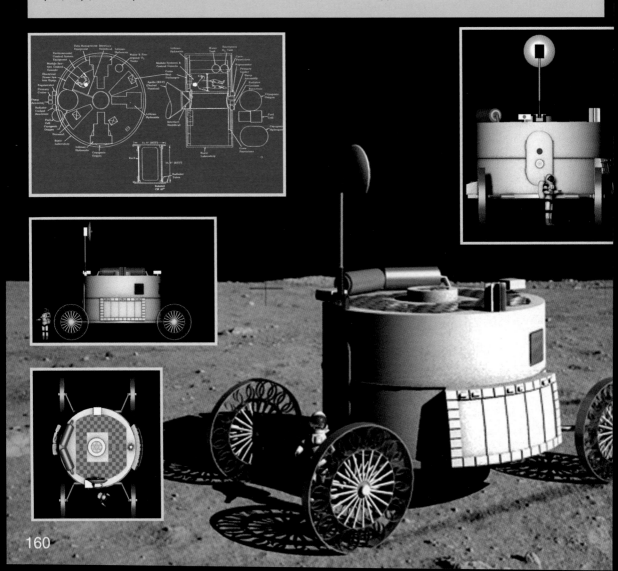

[MOBEV (Continued) 1966]

In the MOBEV studies we are introduced to the acronyms used to define the vehicles.

ROVING VEHICLES
1. First (letter) (R) defines vehicle as Rover.
2. Second (number) (0 through 3) defines vehicle crew size.
3. Third (letter) (A through D) defines specific mission of vehicle.
4. Fourth (letter) (E or B) defines vehicle as being Exploration or Base Support vehicle.
Example: R1BE (rover—one man—vehicle B mission— exploration vehicle)

FLYING VEHICLES
1. First (letter) (F) defines vehicle as Flyer.
2. Second (number) (0 through 3) defines vehicle crew size.
3. Third (letter) (A through E) defines specific mission of vehicle.
Example: F1B (flyer—one man—vehicle B mission)

This rather handy nomenclature allowed the reader to quickly determine the purpose of a specific design.

The report then went on to outline 26 different roving vehicles, as follows:

EXPLORATION VEHICLES

R0AE	SLRV 62.1 kg total 4kg payload
R0BE	Pack Mule 170 kg total,
R0CE	RUNT (Remote Unmanned Traverser) 50 kg payload 148 kg total
R0DE	RECONE (Remote Controlled Explorer)
R1AE	Go-Cart 288 kg total
R1A(1)E	Pack Mule or as a Go-Cart 75 kg payload and 142 kg payload or driver 379 kg total
R1BE	One-Man LSSM 864 kg total, 320 kg payload
R1B(1)E	LSSM with remote control 320 kg payload, 1272kg total
R1CE	SITE (Scientific Instrument Traverse Explorer)
R1DE	One man cabined LSSM 320 kg payload 1935kg total
R2AE	Two- Man LSSM
R2BE	8d MOLAB
R2C(1)E	14d MOLAB 320 kg payload 3398 kg total
R2C(2)E	MOLEM
R2C(3)E	MOCOM
R2C(4)E	MOCAN
R2C(5)E	14d MOLAB (*Boeing*)
R2DE	28d MOLAB
R3AE	28d MOBEX 5557 kg total 700 kg payload
R3BE	42d MOBEX 7646 kg total 1500 kg payload
R3CE	14d MOLAB 3826 kg total 320 kg payload
R3DE	90d MOBEX 8344 kg total 1500 kg payload
R4AE	56d MOBEX

BASE SUPPORT VEHICLES

R0AB	LSSM Trailer 650 kg payload (R1B1E flatbed chassis) 1272 kg total
R0BB	MOLAB Trailer 2798 kg payload (R2C1E flatbed chassis) 3398 kg total
R0CB	MOBEX Trailer 6441 kg payload (R3BE flatbed chassis) 7646 kg total
R1AB	LSSM Prime Mover (R1B1E with digger) 300 kg payload 1272 kg total
R1BB	SITE Prime Mover (R1CE with digger) 300 kg payload 1935 kg total
R1CB	Lunar Tractor 1809 kg payload plus digger 4539 kg total
R2AB	8d MOLAB Prime Mover
R2BB	14d MOLAB Prime Mover (R2C1E with digger) 290 kg payload 3398 kg total

LSSM	Local (Lunar) Scientific Survey Module.
SLRV	Surveyor Lunar Roving Vehicle
MOLAB	Mobile Laboratory
MOBEX	Mobile Excursion
Prime Mover	Base support vehicle, usually with less range and a backhoe.

The SLRV was originally developed in 1964 by *Bendix* and GM for JPL. At one point the SLRV was considered for an advanced version of the Surveyor robotic lander program. MOLAB was an all encompassing acronym for pressurized mobile laboratories, and finally, MOBEX represented a long duration Mobile vehicle i.e. Mobile Extension. The different categories of roving vehicle included:

Unmanned Probe Concepts	for basic remote exploration.
Minimum Manned Vehicle Concepts	similar to what actually flew to the moon in 1971.
LSSM Concepts	larger rovers with higher payload capacity.
One-man Cabin Concepts	small pressurized vehicles.
MOLAB Concepts	large pressurized vehicles.
MOBEX Concepts	long duration large pressurized vehicles.

The vehicles selected for further study were:

R0AE,	SLRV with an RTG power source and range of 36 km
R0BE,	Pack mule with range of 36 km
R0CE,	Remote unmanned traverser with a range of 200 km
R0AB,	Greater versatility LSSM trailer
R0BB,	MOLAB trailer
R0CB,	MOBEX trailer
R1BE,	Baseline LSSM with a range of 360 km
R1A(1)E,	Greater versatility Go-cart with a range of 144 km
R1DE,	Cabined LSSM with a range of 500 km
R1B(1)E,	Greater versatility LSSM with a range of 360 km
R1AE,	Go-Cart with range of 240 km
R1CB,	Lunar tractor capable of 15 km per day
R1AB,	Greater versatility LSSM prime mover 35 km per sortie
R1BB,	Cabined LSSM prime mover 500 km per sortie
R2BB,	14 day MOLAB Prime mover 400 km per sortie
R2C(1)E,	14 day MOLAB with a range of 400 km
R3AE,	28 day MOBEX with 800 km range
R3BE,	42 day MOBEX with a range of 1600 km
R3CE,	3 man MOLAB with 400 km range
R3DE,	90 day MOBEX with a range of 3425 km

The MOBEV studies operated on the basis that a series of possible lunar missions be defined and then a series of vehicles be designed to fulfill those missions. It was believed that the primary vehicles chosen for this would be unique, with an adequate coverage of mission requirements, an evolutionary growth potential, capable of being delivered to the moon by existing hardware and powered by subsystems that would be state of the art. The overall study analyzed over two dozen vehicles before concluding that the overall mission needs could be covered by five exploration vehicles and one base support vehicle, with eleven more as back-up designs, should the need arise.

This wide array of rovers varied considerably in size and potential, consequently each would have been delivered to the moon by the appropriate size of lander. There were five potential delivery systems on the drawing boards. The *LM Shelter*, the *LM Truck*, a modified *Surveyor* and two unmanned vehicles, the *Lunar Logistics Vehicle* (LLV) and the *Lunar Applications of a Spent S-IVB/IU Stage* (LASS).

In the interests of conservation of resources it was decided that a minimum number of exploration rovers could be adapted to become support vehicles for a potential moon base. To make a base support vehicle it was simple to just take an exploration vehicle and equip it with an appropriate heavy duty digger. For moving around large quantities of equipment the basic chassis for the exploration rovers could also be reconfigured as a trailer. The one exception to this seems to have been the R1CB Lunar tractor which was a large stand-alone dump truck/digger/plough with walking beam suspension, this monster would have been taken to the moon aboard the LLV.

All of the pressurized rovers would have expendables based on taking two EVAs per day and would have enough to support the crew for up to seven days. All of the manned rovers, designed to be sent in advance of the landing party, would be created to be able to withstand a six month lunar stay without any failures. Unmanned rovers would not need this storage requirement.

The manned long duration rovers would not carry fuel cells and would need to have water reclamation equipment that could recover 95% of the urine and 98% of the sweat from each crewman. Over two pounds of food per man, per day was supplied.

The exploration rovers that had been converted to base support logistics vehicles would have a backhoe placed on the front with a bucket capacity of just over a quarter of a cubic meter; capable of digging a hole over four meters deep. The backhoe would be used for moving materials, but also potentially for burying a nuclear reactor. Weight restraints would mean that the backhoe would replace the scientific equipment normally placed on board the exploration vehicles, and a trailer hitch would also be added to accommodate the three trailer concepts. The three trailers would have been capable of carrying from 800 to 6400 kg of payload. They would have been based on the chassis of larger vehicles as follows:

R0AB trailer = R1BE chassis (LSSM)
R0BB trailer = R2C(1)E chassis (MOLAB)
R0CB trailer = R3BE chassis (MOBEX)

All of these trailers would have been useless without the base support vehicles to tow them. They were not for taking on long haul exploration and the unmodified explorer rovers would not be equipped with the trailer hitch to accommodate trailers.

The Unmanned Probe Concepts were based on the SLRV designed in 1964 by *Bendix*, GM and JPL.

The medium size unpressurized vehicles were based on the LSSM chassis and the R1BE was actually built in full scale mock-up.

The large, long-duration rovers were powered by fuel cells if they were to undertake missions of less than 30 days. The water load would thus be minimized as the fuel cells would supply surplus water to the crew. On the other hand, longer missions would use one or two SNAP reactors and would have to carry extra water and reclamation equipment

The basic MOLAB system was comprised of the following subsystems: communications, navigation, illumination and TV, and command and control. The communications subsystem provided for S-band communications between the MOLAB and earth, and for local lunar communications at VHF between the MOLAB and other elements, including the command module, the LM, and other possible lunar terminals. The TV subsystem included a TV camera providing a forward-looking stereo pair, one monoptic camera for rear viewing, and three cameras for cabin internal viewing and monitoring. The command and control subsystem provided for the signal switching and distribution required of both telemetry and command signals throughout the vehicle, which was required to implement the manual/remote control modes of operation. The illumination and TV subsystem included a total of seven TV cameras, two of which were gimbal-mounted for azimuth and elevation slewing. The forward and rear-looking antennas were provided with artificial illumination. The navigation subsystem had two modes of operation: position fix and dead-reckoning. The position fix mode utilized a periscopic theodolite along with an inclinometer. The dead-reckoning mode utilized a computer supplied with inputs from the directional and vertical gyros and the odometer. An onboard map display was provided which was driven by the navigation computer and recorded and displayed the dead-reckoning computations.

The smaller rovers with nuclear power used the SNAP 27 reactor, while the larger ones used an actively cooled SNAP 29. Generally the MOLABs and the small pressurized RIDE would have had a pure oxygen atmosphere but the MOBEX would have a mixed gas atmosphere for both safety and health reasons. The one man RIDE would dump its atmosphere for each egress/ingress, while the MOLAB and MOBEX class of rovers would both have airlocks.

Serious thought was given to how much living space would be afforded to each crewman. Anything below 10 cubic feet per person was considered unsatisfactory, between 45 and 90 cubic feet were borderline for missions over five days. Volumes per man of between 100 and 200 cubic feet were considered adequate for missions up to 30 days. The RIDE had a total of 70 cubic feet, the MOLABs about 270 cubic feet and the MOBEXs about 370 cubic feet.

The wheels were extremely important to the design and came in three sizes. 24 inch, 45 inch and the very large 80 inch. The 45 inch wheels were generally for the LSSM while the 80 inch were for the MOLAB and MOBEX. Two different types of wheel were considered; defined as—discrete element metal-elastic (that would be the one with all of the circular springs) and spiral element metal-elastic (that would be the one created by *Grumman* and used on the Mars Rovers decades later.)

Unmanned Probe Concepts

In the initial spectrum, two vehicles were included in this category, R0AE and R0CE. R0AE, the Surveyor Lunar Roving Vehicle (SLRV), resulted from the design studies performed in 1964 and 1965 at JPL. The maximum upper limit for the Surveyor Lander payload launched from earth was assumed at 68 kg (150 lb). Consideration of this fact and the two designs mentioned above had led to the recommendation that an RTG-powered vehicle weighing 68 kg with a limited scientific mission support capability be provided. This vehicle would provide day operation and night survival and an operating duration of 90 days.

[ROAE SLRV 1966]

R0AE was a very minimal vehicle designed for lunar-day-only operation with no night survival. This resulted from the weight constraints coupled with development schedule problems. Very little mission flexibility was inherent in the design. The scientific return would be small, as the primary purpose of the vehicle mission was site certification for manned landings. Later studies showed site certification to be of little value in an overall exploration scheme. An earlier concept of the SLRV was also developed which was powered by an RTG unit and provided night survival and day operation. However, this was still a minimal concept, because it was constrained to remain below a 45-kg (100-lb) limit.

[ROCE RUNT 1966]

The R0CE vehicle was designed to be accommodated by a Surveyor Lander launched from a lunar-orbiting spacecraft. This technique allowed a much higher mass capability up to 175 kg (385 lb) and thus afforded the opportunity for greater mission flexibility. The vehicle designed around this constraint would provide all of the scientific mission support (50-kg payload consisting of a wide variety of surface sampling and analysis instrumentation) capability recommended.

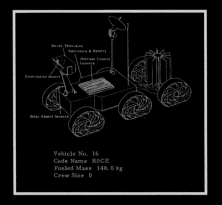
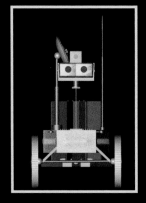
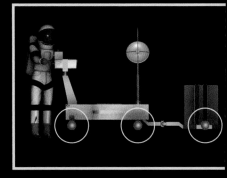
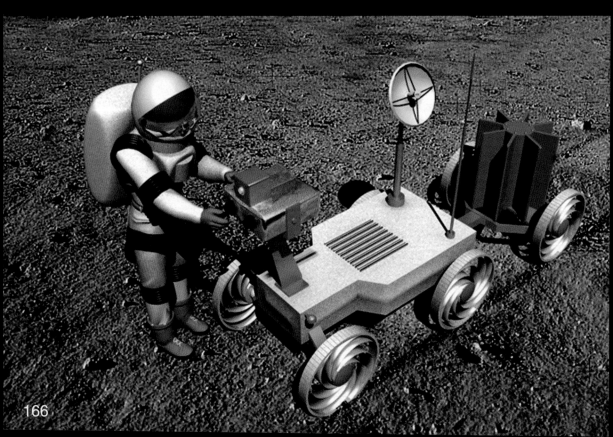

[ROBE Pack Mule 1966]

Minimum Manned Vehicle Concepts

Two vehicles, R0BE and R1AE, were included in the initial spectrum in this category. Each vehicle offered a unique functional capability, and each was in a weight class compatible with early manned lunar missions. R0BE was a powered cargo carrier or wheelbarrow and technically not a manned rover. It was intended to provide a walking astronaut with assistance in the deployment of surface packages normally too heavy (greater than 75 kg) for him to handle. Two concepts for this vehicle were developed. These vehicles in themselves provided little mobility augmentation to a surface mission other than cargo moving which may have saved some time in early deployment operations. However, it was recommended that the R0BE vehicle be retained as a secondary design point by virtue of its unique functional capability and low mass which may have been a critical factor in early mission planning.

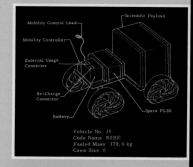

Summer 1966 version of Bendix Pack Mule (right)

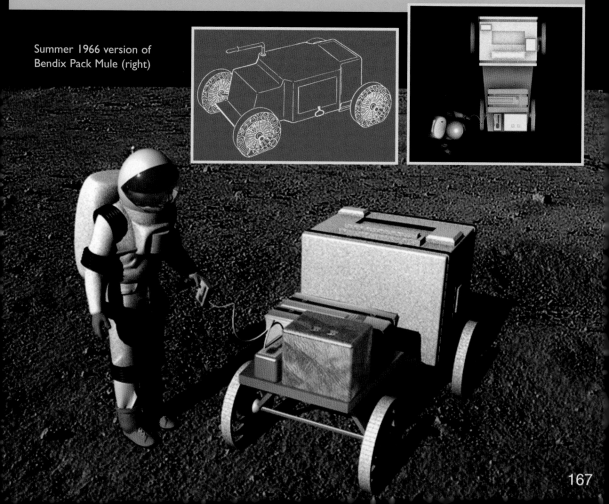

[R1AE Go-Cart 1966]

R1AE was a minimal manned vehicle for limited reconnaissance and sample gathering. It resulted from considerations of what could be done to augment the basic Apollo mission with a mobility capability. It was intended to replace the 114-kg (250-lb) scientific payload of the LM. The maximum achievable total range for the self-contained system was 38 km for no scientific payload capability. For a nominal payload of 10 kg, the mission range was approximately 12 km which would provide valuable reconnaissance and geologic sample-gathering capability. The maximum single sortie range for the rechargeable system was about 19 km. The rechargeable system could be utilized for about 30 sorties. This vehicle was also recommended because of its low mass characteristics and the criticality of this property in early mission planning. Comparative analysis of the vehicles discussed above led to the conclusion that a new concept be developed which combined the functional capabilities into a single vehicle. Since such a vehicle was thought to exceed the payload of the basic Apollo LM, it was recommended that the design be made compatible with the *Augmented LM* concept (a version with more fuel and a stronger structure capable of landing more payload). Therefore, a delivered mass in the order of 300 kg was possible. By including this vehicle an intermediate capability between the minimal vehicles and the LSSM class was provided which more nearly met the needs of Augmented LM missions without undue weight penalties.

[R1A(1)E Go-Cart 1966]

The R1A(1)E was similar in function to the R0BE. Both were used by an ambulating astronaut, increasing his range and equipment-carrying capability. However, the R1A(1)E and the R1AE could, by offloading some scientific payload, transport the astronaut himself, giving additional speed and range. The R1AE was similar to the R1A(1)E in that both carried the astronaut and a payload, but the R1AE's payload was limited to 10 kg. It was realised that the R1A(1)E could be delivered by the standard Apollo Lunar Module. In this respect the R1A(1)E was the nearest thing in the MOBEV studies to what actually flew to the Moon in 1971. It could carry a payload of 75kg as well as having capacity for 142kg for the crew. In all other respects its capabilities were the same as the R1AE.

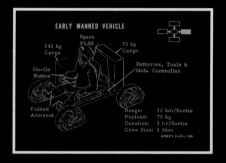
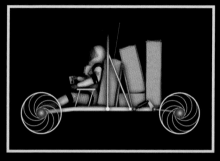
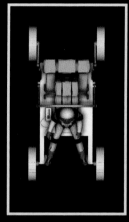

[RODE RECONE 1966]

LSSM Concepts

Several variations of the *Local Scientific Survey Module* (LSSM) were contained in the initial spectrum (R0DE, R1BE, and R2AE). R0DE Remote Controlled Explorer (RECONE) was configured strictly for unmanned use and had a completely self-contained power system consisting of an RTG-battery unit.

[R1BE LSSM 1966]

R1BE and R2AE were one-and two-man versions of the LSSM utilizing a battery power system requiring recharge from an external source. The final 1968 version (inset below), listed simply by Bendix as the *LSSM*, also carried a small RTG over the right rear axle to provide heating for critical components, in all other respects it looked exactly like the 1966 version seen below. All three vehicles provided full astrionics systems for direct earth communications and support of scientific experiments.

The first concept (R1BE) was a "basic" or stripped version completely dependent on a shelter or base for recharge power and thermal control during storage or standby periods. In addition, no astrionics functions (navigation or communications) were retained aboard the vehicle other than that afforded by the PLSS unit and scientific instruments. Curiously the diagrams in the MOBEV studies show the R1BE equipped with a large S-band antenna on the roll-bar and two other antenna for communications. This inconsistency does not seem to be explained in the report. Some scientific power was retained to provide a minimal support capability. Power was supplied by rechargeable batteries. A 320-kg scientific payload was accommodated. The full-size mock-up of R1BE built by *Bendix* also had a large communications array on the roll-bar as seen in the accompanying pictures.

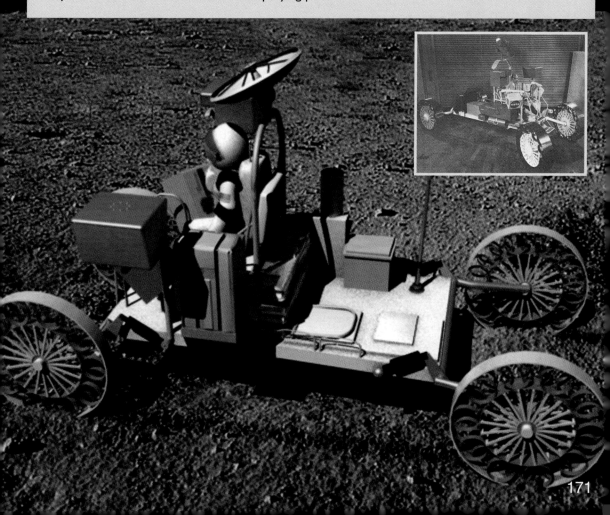

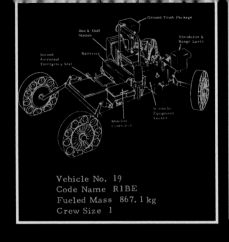

Vehicle No. 19
Code Name R1BE
Fueled Mass 867.1 kg
Crew Size 1

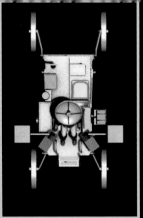

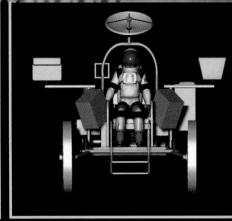

The R1BE LSSM seems to have been one of the few Bendix vehicles built almost as it would have appeared on the moon. (below) The LSSM in a test chamber (inset below)

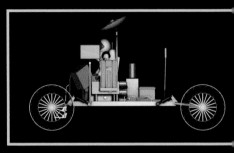

[R2AE LSSM 1966]

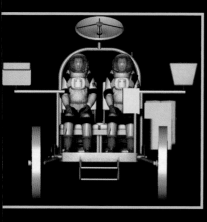
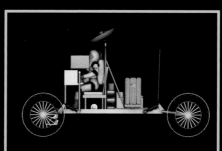
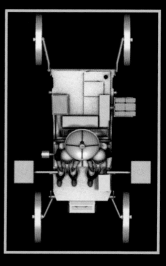

[R1B(1)E LSSM 1966]

The second concept R1B(1)E was a "greater versatility" variation of the earlier versions. It combined all of the mission support capabilities offered in those versions, combined with remote control capability. It was completely independent of the shelter or base. An RTG-battery power unit was utilized. Furthermore, the RTG unit size, 650 w, was based upon reuse in another concept (R1DE) rather than the specific mission requirements. A high degree of mission flexibility was offered by this concept (e.g., 90-day operating duration, and remote control operation prior and subsequent to manned missions). In addition, the larger RTG unit made the vehicle capable of continuous use on a demand basis when the astronaut was off loaded.

[R1CE SITE 1966]

One-Man-Cabin Concepts

As an advanced growth version of the LSSM vehicle, a 12-hr sortie vehicle with a minimum cabin (R1CE) was included in the initial vehicle spectrum. It was referred to as the SITE or *Scientific Instrument Traverse Explorer*. This vehicle provided extended range and scientific time on-station, over the basic LSSM concepts with open cabins. As a result of recommendations by NASA, this design point was shifted to a 48-hr, one-man-cabin vehicle (R1DE). Once again, at least one error appears in the MOBEV studies that misidentifies the large three-man R3DE MOLAB vehicle as the R1DE and vice versa. The primary variations available to the R1DE concept were in the power supply system. Three types of systems were applicable to this mission: a rechargeable battery system which was dependent on a base recharge source, a fuel cell system, and an RTG battery combination.

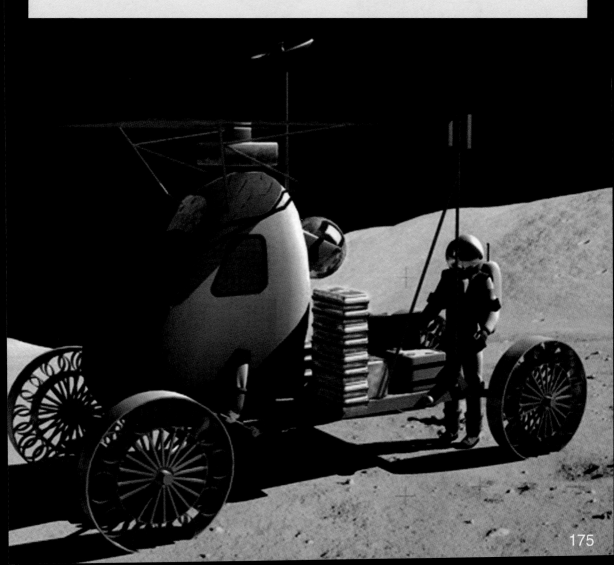

The front hatch on the RICE and RIDE would have had the controls mounted on the inside of the door (above) The vehicle was rated for multiple pressurizations and depressurizations.

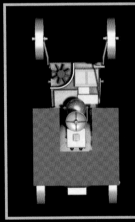

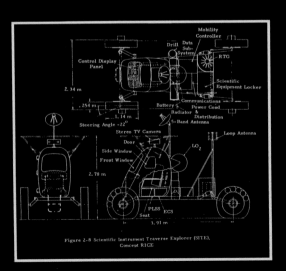

Figure 2-8 Scientific Instrument Traverse Explorer (SITE), Concept RICE

176

[RIDE SITE 1966]

The Cabined LSSM, RIDE, was a manned exploration vehicle designed to provide a shirt-sleeve (open spacesuit faceplate) environment. The vehicle was to be delivered by a *LM-Truck* and was designed for up to six months of lunar storage prior to commencement of the mission. Prior to the crew's arrival, the vehicle was remotely unloaded and driven to the site of the astronaut's landing. The vehicle's mission was to provide three 48-hr sorties, and three 8-hr sorties during the 14 days the crew was on the moon. The range per sortie is 125 km per 48-hr sortie, and 40 km per 8-hr sortie. The vehicle carried a 320 kg scientific payload, and had an average speed of 8 km/hr.

The cabin was a basic semi-monocoque pressure shell. A large front-viewing window was located in the full-length hinged door, and two side windows were also provided to enhance astronaut visibility. The astronaut seat incorporated a restraint system. The basic control and display pedestal was attached to the exit door permitting unobstructed entry and exit from the cabin.

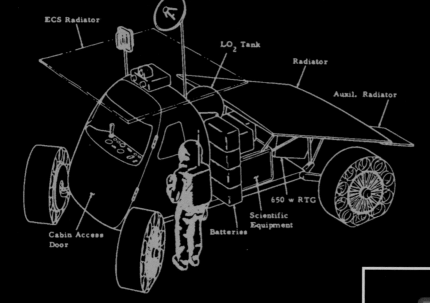

Vehicle No. 21
Code Name R1DE
Fueled Mass 2022.0 kg
Crew Size 1

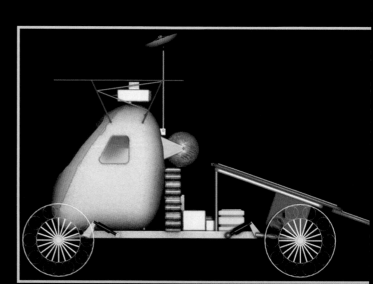

Notice the large radiator panel mounted on the R1DE (not on the R1CE). This allowed increased range.

[R2C(1)E MOLAB 1966]

MOLAB Concepts

MOLAB concepts were two-man-cabin vehicles. There were seven concepts in this category in the initial spectrum, R2BE, R2C(1)E, R2C(2)E (MOLEM), R2C(3)E (MOCOM), R2C(4)E (MOCAN), R2C(5)E and R2DE. Three of the concepts, MOLEM, MOCOM, and MOCAN, were discussed earlier. All of the remaining concepts were basic MOLAB vehicles, R2BE and R2DE representing scaled versions of the MOLAB studied in the *Apollo Logistic Support Systems* payloads studies. The R2BE concept was an 8-day version of the MOLAB. Because of the shorter mission duration, the cabin was shortened and the airlock concept was changed. This concept offered no specific advantages over the basic MOLAB other than mass savings which were not a critical factor since the *LM-Truck* or *LLV* would be

utilized for delivery. The other two concepts, R2C(1)E and R2DE, were derived by direct scaling. Mission extensions beyond 28 days required changes in life support system type (two-gas system) and an increase in cabin free volume. The 14-day MOLAB was selected as the primary design to represent this class of vehicle. A number of mission variations using this vehicle were possible through several alternatives. It was recommended that a 14-day, three-man MOLAB vehicle (R3CE) also be included as a secondary design.

The R2C(1)E was a manned mobile laboratory (MOLAB) used for exploration. The MOLAB provided complete life support capabilities for its two-man crew during a 14-day, 400-km mission. A seven-day life support contingency, beyond the basic 14 days, was also included in the design.

The MOLAB carried 320 kg of scientific payload, and had an average driving speed of 10 km/hr. The MOLAB was delivered to the moon by a *LM-Truck*, and was capable of being stored for as long as six months before commencing its mission. Prior to the crew's arrival, the MOLAB was remotely unloaded and driven to the site of the crew's landing. The 14-day mission commenced with the arrival of the crew.

Outwardly there seems to have been little to distinguish the R2BE, R2DE, R3CE and the R2C(1)E. Therefore only one set of illustrations has been provided here.

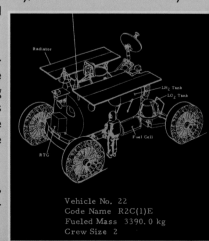

Vehicle No. 22
Code Name R2C(1)E
Fueled Mass 3390.0 kg
Crew Size 2

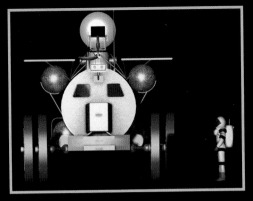

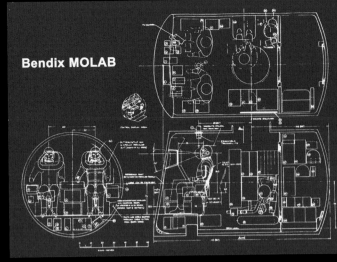

Bendix MOLAB

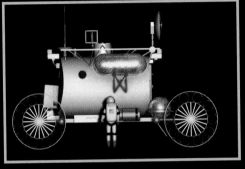

An R2C(1)E desk model deploys (right)

The R2C(1)E was outwardly very similar to the R2BE, R2DE and R3CE.

GM/Boeing MOLAB 944-001 1964

The R2C(5)E is referenced as a design for a MOLAB created by *Boeing*. Details of this *Boeing* design are very limited in the available MOBEV literature. However, it seems likely that the R2C(5)E was somehow related to the 14-day MOLAB 944-001 designed by *Boeing* and *General Motors* that appears here. This vehicle seems to have been built at full-scale and is clearly riding atop the *Boeing* six-wheel flexible frame that had been tested at Aberdeen proving grounds. It appears to have been capable of towing more of the two-wheel "trailer" units to increase its stay time up to as much as 45 days. The full-scale mock-up and the few diagrams that exist of the *Boeing/GM* MOLAB included an air-lock which may have been equipped with a docking adapter to accommodate retraction from the SLA by an Apollo CSM. It was also to have landed aboard a *LM Truck (bottom right)*. The trailer carried LH2 and LO2, batteries, fuel cells and radiators. The vehicle had 300 cubic feet of living space. The vehicle and 2 man crew weighed 6400 lbs and was just over 20 feet long.

The GM/Boeing MOLAB mock-up (right)

Testing rear hatch (right center)

Mounted on ILMT for landing (bottom right)

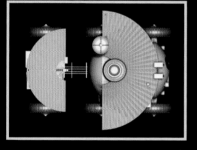

The 944-001 MOLAB came at the end of a long line of studies by Boeing undertaken during the spring and summer of 1964. Before the final design was accepted an assortment of configurations were investigated. The 944-001 was a six wheel semi-articulated vehicle but other configurations included: six wheel articulated, four wheel articulated and four wheel rigid frame. Various different cabin shapes were also considered including: horizontal cylinder, rectangular and vertical cylinder.

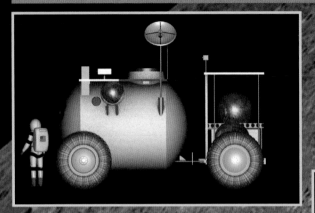

Very similar to the final configuration was the four wheel articulated MOLAB concept. Weighing in at 7080 lbs and 191 inches in length, this vehicle had a 120 inch wheelbase and with the top radiator was 145 inches wide. The tread was 110 inches. The cabin was horizontal cylindrical and contained 300 cubic feet of habitable volume. The trailer was powered and carried LO2 and LH2 supplies.

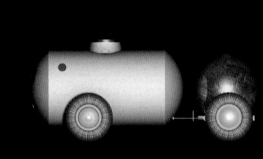

A horizontal cylindrical cabin was also used for model 25-52104 with a weight of 6935 pounds. Articulated steering with ±30° capability used the articulation joint just aft of the crew compartment. Although not shown here it was to have included radiators for both modules.

A rectangular cabin was used for model 25-52103 (right) with a weight of 7814 pounds. Again external power sources were mounted on the rear trailer.

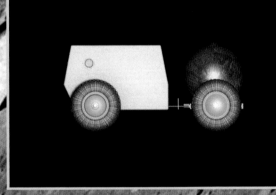

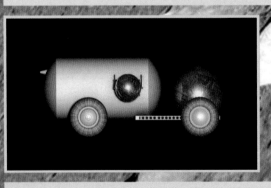

A rigid frame was used for model LO2-2518-16B (above). The same frame was also considered for rectangular and vertical cylindrical models (derived from 25-52105 and 25-52103)

A vertical cylindrical cabin was used for model 25-52105 (bottom right) with a weight of 7032 pounds. As with all of these models the air lock was located in the aft end of the crew compartment.

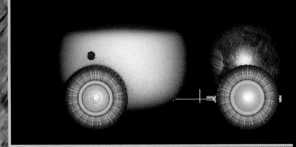

Model 25-52122 was a fully articulated six wheel vehicle weighing in at 7335 pounds. Spanning almost 25 feet in length and 11 feet wide it had articulated steering at both front and back. The horizontal cylindrical cabin was mounted sideways on two wheels. Each wheel was individually suspended and powered. Again 300 cubic feet of space was available to the crew, including the airlock. The front module carried an RTG for power.

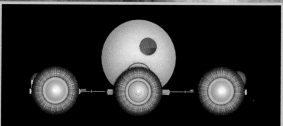

Model 25-52107 was a fully articulated six wheel vehicle weighing in at 7395 pounds. Hub mounted motors provided locomotion. Both front and rear units contained cryogenic tanks and batteries. A version with a vertically mounted cylinder was considered but could not be accommodated by the Saturn V SLA.

Model 25-52115 was a semi-articulated six wheel vehicle weighing in at 7019 pounds. The crew compartment is a horizontal cylinder with an elliptical cross section and domed ends. The driver is located at the front and airlock at the rear.

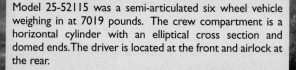

Model 25-52106 was a fully articulated six wheel vehicle weighing in at 8229 pounds. The only major difference between this and Model 25-52107 was the shape of the cabin. Antennas, tv cameras, environmental support controls and radiators would all have been attached to the crew compartment. (*contractor model above*) (*1965 Westinghouse moon base model above right*)

Model 25-52108 was a semi-articulated six wheel vehicle weighing in at 7106 pounds. This model employed a compartment comprised of two vertical cylinders. The front cylinder is both the driving compartment and the airlock. The rear wheels were deployed on a parallel set of flexible frames that provided pitch and roll articulation between the 4 by 4 front section and the 2 by 2 rear section.

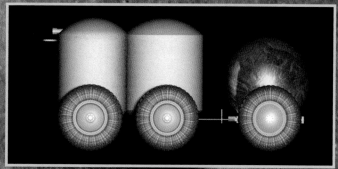

[R3AE MOBEX 1966]

MOBEX Mobile Excursion Laboratory Concepts

This category of vehicles included those with larger than two-man crews and mission duration extending beyond 28 days. The need for such vehicles in extended lunar exploration programs had been identified in several places. There were three vehicles in this group in the original spectrum, R3AE, R3BE, and R4AE. R3AE and R3BE provided three-man crews with exploration capabilities of 28 and 42 days, respectively. As in the MOLAB case, both vehicles utilized the same basic cabin structure. This cabin provided a free volume of 372 cu ft. This did not meet the maximum volume requirement of 150 cu ft per man. However, in view of the fact that ample opportunity was provided for EVA, this cabin would suffice for mission duration of any practical length (90 days would meet all postulated mission requirements). R3BE and R3AE appeared superficially the same.

[R3DE MOBEX 1966]

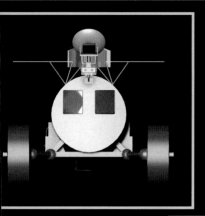
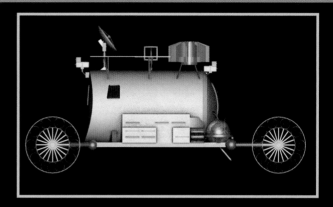
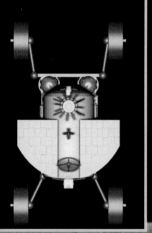

The 90-Day MOBEX, R3DE was a manned mobile laboratory used for exploration of the moon. The vehicle provided complete life support capabilities for its 3-man crew during a 90-day, 3425-km mission. The vehicle carried 1500 kg of scientific payload, and had an average driving speed of 10 km/hr. The R3DE was delivered to the moon by an LLV and was capable of being stored for as long as six months before commencing its mission.

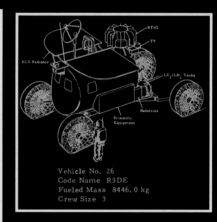

Vehicle No. 26
Code Name R3DE
Fueled Mass 8446.0 kg
Crew Size 3

[R4AE MOBEX 1966]

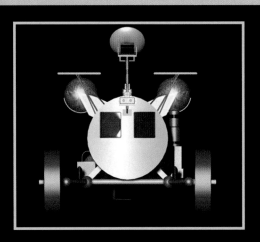

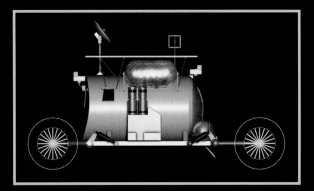

Bendix contractor models (opposite) show slightly different configurations to the MOBEV studies (note the round radiator)

R4AE, a four-man vehicle with a duration of 56 days and a range of 2000 km, was dropped from further consideration in the vehicle spectrum because of incompatibility with the most likely crew delivery schemes that provided for delivery of three men or multiples thereof (i.e. Apollo.)

The primary trade-offs in the MOBEX class of vehicle were concerned with the particular power and life support system types to be utilized. Three power systems were considered: fuel cells, isotope Rankine, and isotope thermionic systems. The fuel cell systems provided life support oxygen storage, and the water produced in the fuel cell operation was also utilized. The significant changes in life support resulted from water management studies. Without potable water from the fuel cell for drinking and PLSS (EVA backpack) needs, water management became a major problem. A water reclamation system could be used to recover approximately 98% of liquid waste. This, however, could only meet drinking water requirements. PLSS cooling water would have to be stored. The specific advantages of the fuel cell systems included earlier availability of components, elimination of radiation-interference problems with scientific experiments such as photographic equipment, and elimination of radiation hazards to the crew.

The isotope systems offered the advantage of a lower delivered mass and a greater mission growth potential for extended duration missions of 90 days or more, since the only requirements would be for additional life support expendables. The disadvantages of the fuel cell systems were the highest delivered mass and a limited growth capability due to the excessive tankage mass and sizes required for extended duration missions. The isotope systems required advanced component development. They also introduced complex integration problems associated with thermal control and radiation shielding from both the crew and the sensitive scientific equipment. As a result of these considerations, and the desirability of having a long-range vehicle contained in the final spectrum, it was recommended that R3DE (a 90-day, three-man, isotope-powered vehicle) be designated as a primary design. R3BE, a 42-day, fuel-cell-powered vehicle, was also included in the primary design category because of its earlier availability and lower development risk potential. R3AE was retained as a secondary vehicle to provide data for an additional design point.

[Northrop WBS MOLAB 1966]

Northrop would continue to compete for MOLAB long after the contract had been awarded to Bendix and GM/Boeing. The Northrop WBS was a 7000 pound/2-man vehicle with eight wheels, two airlocks and what was called Walking Beam Suspension that allowed for unprecedented mobility. Pictured below is a mockup of the Northrop WBS. The walking beam suspension was not a new idea and had previously been used for construction vehicles and railroad cars, however it was a known quantity and considered reliable. The Northrop WBS could have been delivered aboard a LM Truck.

[NASA MOLAB] 1963-1964

The original MOLAB was a joint study undertaken by the MSFC Aero-Astrodynamics Lab, Hayes and Northrop. It was part of the Apollo Logistic Support System (ALSS) and was to be delivered by LM Truck. It weighed 7000 pounds and could house two crew for 14 days. A full-size mock-up was built at MSFC in 1964 (below and page 203). The pictures here show NASA engineer Joseph DeFries (who was in charge of lunar lander and MOLAB studies) with a large ILMT model and the NASA MOLAB. This MOLAB was to work in conjunction with the Bell Hopper (LFV) (see page 198).

[R1AB Prime Mover 1966]

The R1AB prime mover was the R1B(1)E with a digger attached. The extra weight was compensated for by removing consumables and experiments. It had a trailer hitch attached for towing the R0AB.

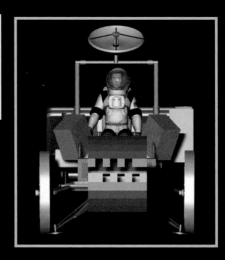

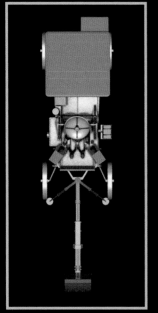

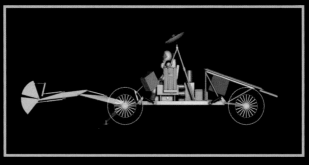

[R1BB Prime Mover 1966]

The R1BB prime mover was the R1CE with a digger attached. The extra weight was compensated for by removing consumables and experiments. It had a trailer hitch attached for towing the R0AB.

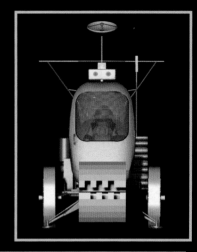

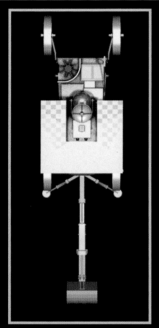

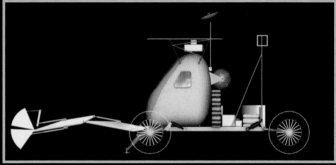

[R1BB Prime Mover (version 2) 1966]

The R1BB prime mover version 2 was the RIDE with a digger attached. The extra weight was compensated for by removing consumables and experiments. It also had a trailer hitch attached for towing the R0AB. It seems that this one makes little sense since the Prime Movers were short duration base support vehicles and the RIDE was designed for longer duration exploration.

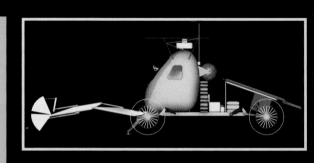

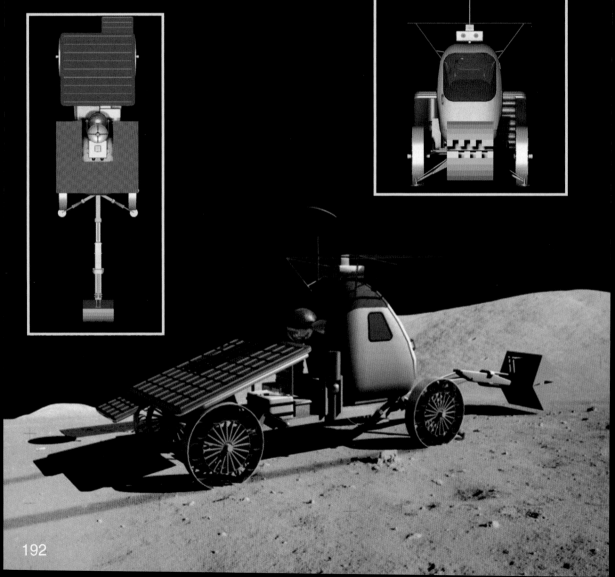

[R1CB Tractor 1966]

The R1CB prime mover was a totally unique heavy base support vehicle which would have required a special LLV or LASS delivery system. Of all of the Prime Movers proposed under MOBEV this seems the most ambitious. The concept of heavy construction equipment had been entertained as early as 1963 by the Martin company. The three photographs at bottom right show Martin's Lunar Transport Vehicle and heavy diggers.

[R2BB Prime Mover 1966]

The R2BB prime mover was the R2C(1)E MOLAB with a digger attached. The extra weight was compensated for by removing consumables and experiments. It had a trailer hitch attached for towing the R0BB large trailer.

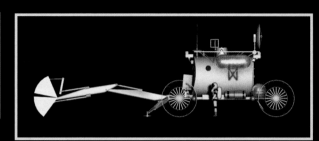
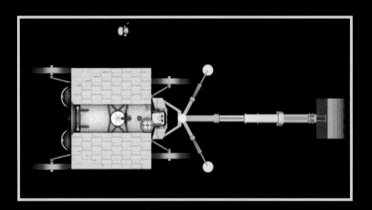
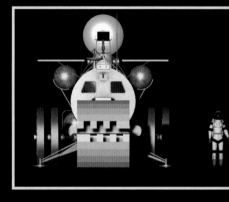

[R0AB Trailer 1966]

The R0AB trailer was an R1B(1)E chassis. It had a carrying capacity of 650 kilograms. It could only be towed by a Prime Mover since the Exploration vehicles did not have trailer hitches.

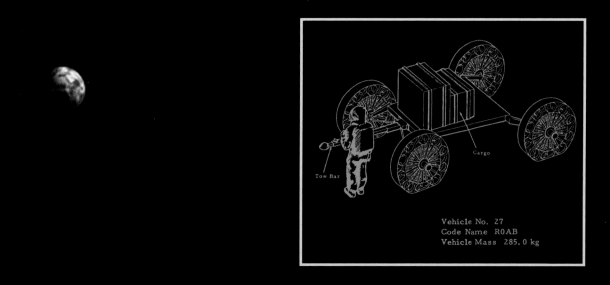

Vehicle No. 27
Code Name R0AB
Vehicle Mass 285.0 kg

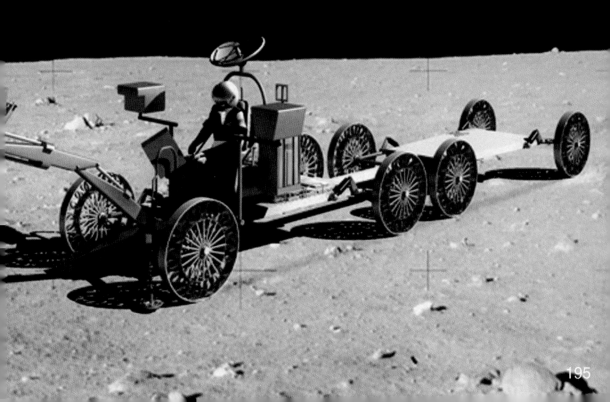

[ROBB Trailer 1966]

The R0BB trailer was an R2C(1)E MOLAB chassis. It had a carrying capacity of 2798 kilograms. It could only be towed by a Prime Mover since the Exploration vehicles did not have trailer hitches.

Vehicle No. 28
Code Name R0BB
Vehicle Mass 565.0 kg

[ROCB Trailer 1966]

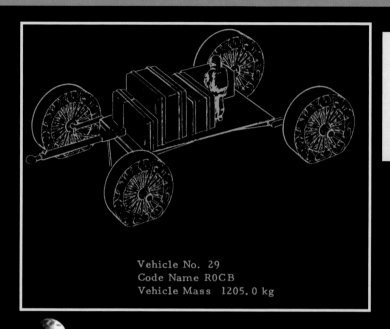

Vehicle No. 29
Code Name R0CB
Vehicle Mass 1205.0 kg

The R0CB trailer was an R3BE MOBEX chassis. It had a carrying capacity of 6441 kilograms. It could only be towed by a Prime Mover since the Exploration vehicles did not have trailer hitches.

[Bell ALSS Hopper 1964]

Flying Vehicles

Bell engineer Wendell Moore had been dreaming of building a one-man rocket backpack since 1953, when he was working on the X-1 rocket plane. Bell began pumping money into his idea in 1957 and early test flights began later that same year. His early design was specifically aimed at providing battlefield mobility to soldiers here on the Earth. Thiokol were also testing similar technology for the same purpose as early as 1958. Moore filed his patent in 1960 and filed further patents in 1961, 1964 and 1966. In his original patent application Moore cited a 1945 French patent for a rocket backpack for skiers. That same patent cited a Hugo Gernsback science fiction magazine from 1931 as the inspiration for the whole idea.

Bell soon became the de facto keeper of the lunar flying vehicle. Their experience with vertical take-off and controlled flight made Bell's engineers the perfect candidates for this task. While *Bendix*, *Boeing* and *Grumman* concentrated on rovers, *Bell* studied the Lunar Flying Vehicle (LFV). A $199,000 feasibility contract was issued by NASA in July 1964. The contract was issued to *Bell* in October.

Toward the end of 1964 the concepts for the *LM-Truck* and Advanced Lunar Modules were already taking hold. This program was called the Apollo Logistic Support System (ALSS). It was determined that a suitable large facility was needed to test these new systems in simulated lunar conditions, right here on the Earth. A variety of existing facilities were considered until in December of 1964 the US Army engineers submitted their final recommendations. The two vehicles used for these studies were the large *Grumman* capsule-shaped MOLAB, with trailers, and a design for a *Bell* lunar flying vehicle which they called the lunar "hopper". This flying vehicle had a dry weight of 500 pounds with a range of 50 miles. It was to be powered by LM propellant, N2O4 and UDMH (unsymmetrical dimethyl hydrazine). It stood 83 inches wide and 70.5 inches high and could carry either two crew or one crewman with a scientific payload. It was not designed to fly into orbit.

An early Bell illustration showing the Hopper and the original MSFC MOLAB in the background (see page 189 for details)

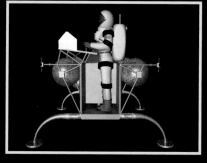
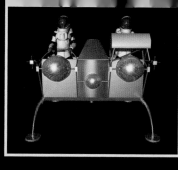
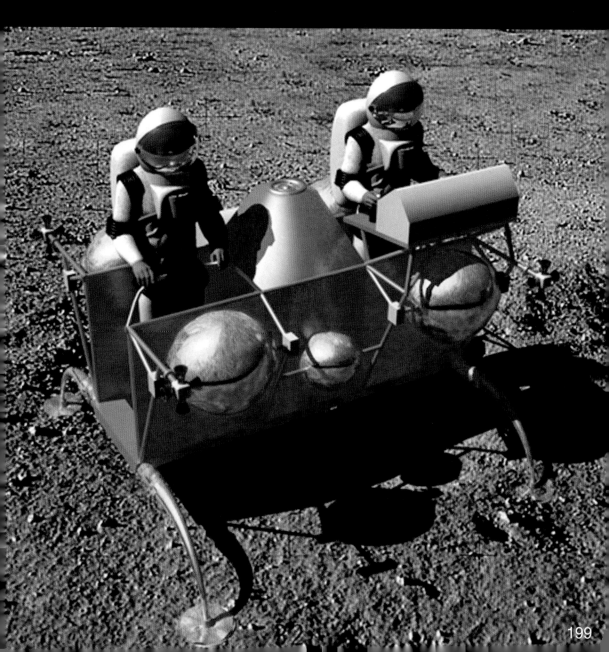

[Bell One-man Escape 1966]

In January of 1966 *Bell Aerosystems Company* submitted a lengthy report called "A Study of Personnel Propulsion Devices for use in the Vicinity of the Moon." This multi-volume study began with these words:

"This report presents results of studies to establish conceptual configurations of propulsion devices which can be used for transportation on the moon, and for escape from the surface of the moon and injection into the lunar orbit. The study was directed toward "simple" devices which make maximum utilization of the perceptual and control abilities of the pilot and minimum automatic flight control and guidance equipment."

In this study the *Bell* engineers presented designs for no less than fourteen different one-man and two-man flying vehicles (see page 212). Most of the two-man versions were similar to the proposed "hopper" from the 1964 US Army report. The one-man escape vehicle seen here was a two stage vehicle. The first stage was a solid rocket motor with 5000 fps delta-v. The center mounted solid rocket would then be ejected and the vehicle would pitch through 90 degrees as the rear mounted hypergolic fueled second engine fired.

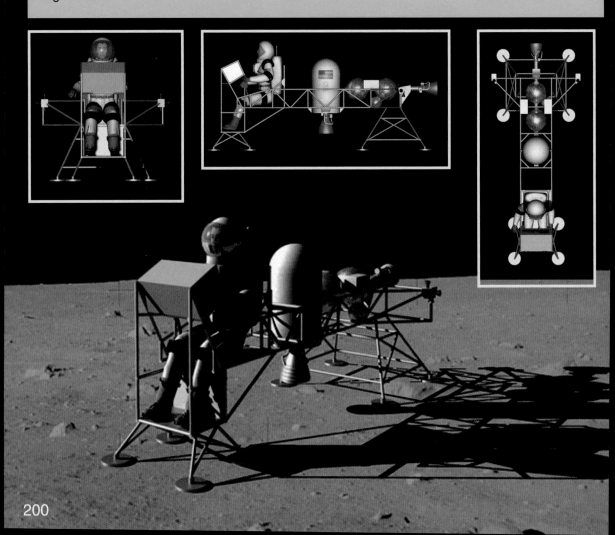

[Bell Rocket Pack 1966]

When *Bendix* handed in the final MOBEV studies in November of 1966, *Bell* had reduced the potential candidates for lunar flying vehicles down to a mere handful. The original *James Bond* style of rocket pack had been revised considerably for lunar use. While the Earth-bound version was powered by a monopropellant (hydrogen peroxide) the lunar version was to have used the same fuels as the lunar module (i.e. N2O4 and 50% UDMH). These rocket fuels were more powerful than the hydrogen peroxide, however they were just as volatile and presented other handling hazards.

Two versions of this adapted model were considered; one with a 115 lb thrust engine and a 2000 fps delta-v while the other had larger tanks and a 163 lb thrust engine with a 4000 fps delta-v. Both were pressurized with gaseous helium and both had enclosures (unlike the terrestrial version) that surrounded the tanks on the astronaut's back. *Bell* had already developed considerable expertise using the terrestrial "small rocket lift device" and they concluded that this backpack mounted system offered substantial improvements over the proposed platform type systems. The whole package included the back-mounted propulsion system and a chest-mounted environmental control system. A small payload may also have been carried on the astronaut's chest but the entire system was rated for a 200 pound astronaut and no more than a 30 pound payload. The backpack employed three propellant tanks to maintain symmetry and alleviate center of gravity excursions. Two throttleable chambers would provide thrust for translation and attitude control. The chambers were mounted on a tubular structure which was pivoted by two swivel fittings on the upper end of the backpack structure. This tubular structure was placed on a torsion bar which permitted the astronaut to apply deflection by pushing up and down on the armrests. When the astronaut released the pressure the tubular arm would return to its détente position. The chest-mounted environment control supplied up to four hours of life-support. Of course the biggest advantage of using a backpack was that you didn't need landing gear. It also had no guidance system because it was only to be used in the proximity of the landing site or moon base.

The Bell rocket pack was a derivative of the terrestrial version seen at right and patented in 1962.

Below are the drawings for two versions of the lunar rocket pack.

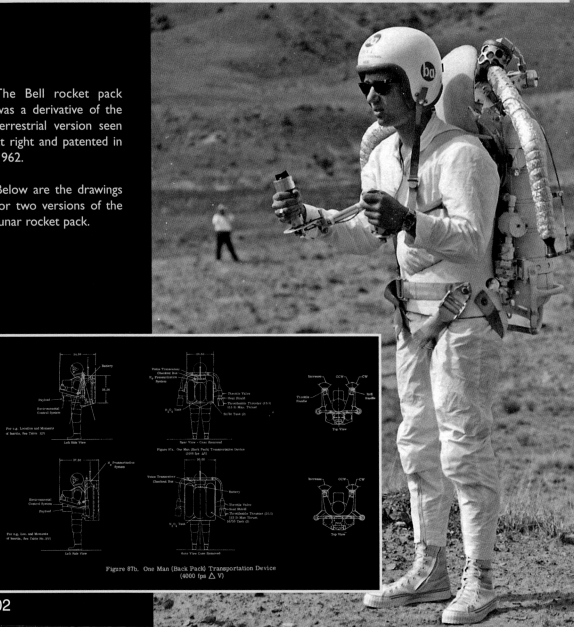

Figure 87b. One Man (Back Pack) Transportation Device (4000 fps Δ V)

202

Flying vehicles could also be used for exploration or rescue. Exploration vehicles were either the single crew, POGO-type configuration (and in one instance a two-man POGO), or were multi-manned vehicles capable of transporting cargo in addition to crew members. Rescue vehicles consisted of surface to surface or return-to-orbit vehicles for rescue of stranded astronauts.

The flyers investigated fell into three types: (1) one-man pogos, (2) multi-man surface-to-surface vehicles, and (3) multi-man return-to-orbit vehicles. One of each type were selected as primary designs. These were the F1B, F2B, and F2E concepts.

Space News announcing the MOLAB and Bell LFV contracts. Note the full size MOLAB mockup pictured (right) see page 189 for details.

The original 1945 design for a rocket backpack for skiers cited by Wendell Moore. (below right) and the Gernsback magazine that started the whole rocket backpack craze (below).

OCTOBER 28, 1964   PAGE 7

Lunar Flying Vehicle And Mobile Lab For Moon Explorers, Subject Of Studies

A rocket-powered vehicle which would enable astronauts to launch themselves from the moon's surface in an emergency and rendezvous with a lunar orbiting craft is one of several lunar transportation devices being investigated by Textron's Bell Aerosystems Company for the National Aeronautics and Space Administration.

Bell has been awarded a contract to provide NASA's Langley Research Center, Hampton, Va., with parametric data on the performance of several types of transportation devices to be used in the vicinity of the moon. This information will be used at Langley for simulation of these various devices in preparation for future space programs.

The variety of vehicles being investigated ranges from a back-mounted device for a single astronaut to a one or two-man platform type configuration. Emphasis will be placed on simple, minimum weight systems.

In addition to establishing performance characterisitics for the vehicles, Bell is also providing conceptual designs for each system.

"For short range missions it appears that a back-mounted device, similar to Bell's Rocket Belt, will be satisfactory," explained technical director Dr. Leonard M. Seal. "For surface translation missions greater than approximately 20 miles, a vehicle on which an astronaut is positioned is required." He added. "The purpose of the vehicle naturally will determine its configuration."

Seal, who is chief of Bell's space systems advanced design section, pointed out that with a rocket-type system an astronaut would be able to fly up the side of a lunar crater or over surface crevasses and explore otherwise inaccessible terrain. In an emergency, the device would be able to carry the astronaut to a lunar orbiting space vehicle, such as the Apollo spacecraft.

Bell recently announced receipt of another NASA contract to conduct preliminary design studies of rocket propelled two-man Lunar Flying Vehicle (LFV) for lunar rescue and reconnaissance missions.

The LFV would be employed in conjunction with a Mobile Laboratory (MOLAB).

The MOLAB concept is being studied by NASA to support a possible 14-day manned lunar exploration mission following project Apollo. The LFV will be designed to enable rapid transit of the astronauts to safety from any lunar danger encountered or from a malfunctioning MOLAB. It also may be used to explore terrain inaccessible to the MOLAB or carry an astronaut on reconnaissance, surveying, photography or mapping missions.

In this concept of the LFV a rocket engine is located between the astronauts. This throttleable propulsion system will enable the LFV to hover, or fly a flight path much like a helicopter. In addition to a single engine vehicle, Bell is providing NASA with preliminary designs of multiple engine configurations.

Reaction control rockets, at the corners of the vehicle above the landing struts, will help maintain the craft's stability in flight. They are powered by the same propellants as the main engine.

Positive expulsion fuel tanks are located beneath the astronauts control consoles and behind the vehicle. Life support and communication equipment is carried on the backs of the astronauts.

MOON EXPLORATION CAR—This is a full size model of a mobile laboratory (MOLAB) configuration being studied by NASA's Marshall Space Flight Center to support roving lunar explorers for up to two weeks. MOLAB would replace the ascent stage of the Apollo Lunar Excursion Module and be placed on the moon by a Saturn V launch vehicle. A second Saturn V launch would send two astronauts to the moon to board the MOLAB.

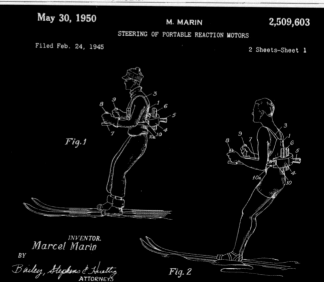

203

[Bell F1A 1966]

The F1A was one of the first POGO designs to come from Wendell Moore's desk. It had a range of up to 8 kilometers but it would be surpassed by the upgraded F1B. It continued to be tested well into 1967.

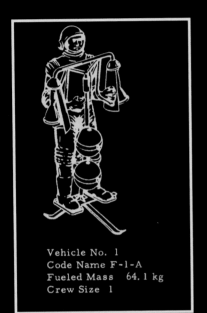

Vehicle No. 1
Code Name F-1-A
Fueled Mass 64.1 kg
Crew Size 1

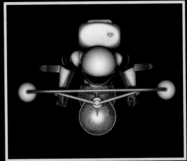

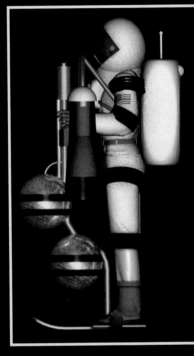

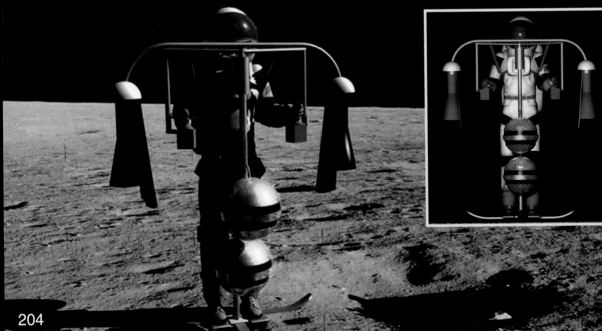

[Bell F1B 1966]

The one-man flying vehicle, F1B, relied on manual control obtained by thruster gimbaling for pitch and roll, and jetavators for yaw. Communication was limited to the PLSS VHF, line of sight, voice system. Navigation and guidance was done visually with aids such as maps and a sextant to determine initial flight direction. The F1B had a potential range of 20 kilometers. Moore filed his patent for this device in 1966 and he portrayed it as an improvement over his earlier backpack design. In this revised patent Moore presented several versions including one which used a turbine engine for thrust (obviously impractical for use on the moon) as well as both standing and seated versions that used hydrogen peroxide fueled rockets.

The seated version was never considered as part of the MOBEV studies.

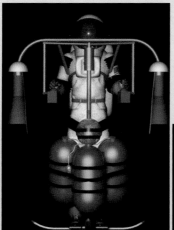

Vehicle No. 2
Code Name F-1-B
Fueled Mass 82.0 kg
Crew Size 1

Inventor Wendell Moore is seen with something similar to the F1B (above left)

[Bell F1C 1966]

The F1C platform flying vehicle was similar to several early *Bell* platform POGO flyers. This one had a range of 170 kilometers and was expected to cost over $13 million to develop.

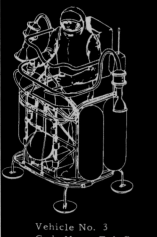

Vehicle No. 3
Code Name F-1-C
Fueled Mass 309.9 kg
Crew Size 1

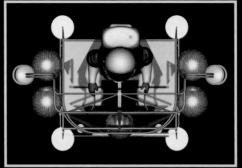

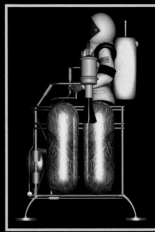

[Bell F2B 1966]

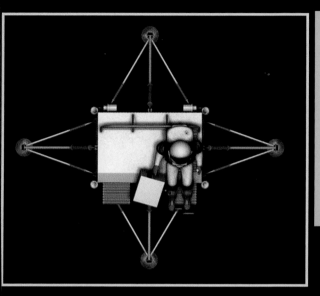

The multi-man surface-to-surface vehicle, F2B, used differential throttling of the lift thrusters. Eight thrusters were used to provide engine-out capability and provide an initial lunar thrust-to-weight ratio of 3. Two thrusters were mounted at the corners of the body in a slightly canted position to provide yaw control. An active three-axis attitude control system was provided. Continuous communications with the lunar base or roving vehicle during exploration was provided by an S-band system integrated with the PLSS system.

Vehicle No's	4	5	6	7
Code Name	F-2-A	F-2-B	F-2-C	F-2-D
Fueled Mass (kg)	287.9	391.8	530.5	788.3
Crew Size	2	2	2	2

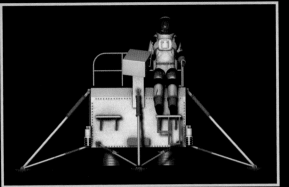

Multiple landing capability was provided by reusable friction devices for landing energy absorption. A strap-down inertial guidance system provided the required guidance accuracy. The F2B had a maximum range of 50 kilometers. A very similar vehicle to the Bell F2B was submitted for a patent by designer William Collins on behalf of NASA in 1970.

A full size mock-up of the F2B

[Bell F2E 1966]

The return to orbit vehicle, F2E, was selected as a primary design. It was provided with six degree of freedom control for rendezvous as well as normal attitude control. Four main lift engines were used for boost and midcourse corrections and positive expulsion tanks were provided. Line of sight communications with the orbiting CSM was provided by the PLSS VHF system, with the addition of an amplifier to increase the range. The navigation and guidance system was the same as the surface-to-surface vehicle with the addition of a rendezvous transponder. Less powerful versions were almost identical in appearance. The F2C had a range of 100 kilometers while the F2D could travel up to 200 kilometers. Three-man versions were also considered as F3A, F3B, F3C and F3D with ranges of 50, 200, 400 and 800 kilometers respectively. The F3E was a return to orbit vehicle. Superficially these vehicles would all have appeared almost identical. In Bendix engineer, Richard Wong's paper in August 1966 the Bell F2E had a wide tripod landing gear, but it seems to have been dropped by the time the MOBEV study was published in November.

In the flying vehicle conceptual designs, the approach was to make maximum use of data developed in past and current programs and to use common concepts for the various vehicles insofar as possible. Thus, the vehicles, particularly those for two and three-man crews, had many similarities such as the same landing gear and structural approach, the same type of propulsion systems using the same propellant, similar configurations in shape, same number and arrangement of propellant tanks, etc. For the initial designs during the conceptual design period certain assumptions were used. For example, thrusters were scaled to suit each vehicle and the multi-man vehicles were all assigned the same guidance system mass. When utilizing a flying vehicle for exploration, it would undoubtedly be used for multiple sorties and although the design point vehicles were capable of performing multiple sorties the vehicles were sized for a single flight of the range specified.

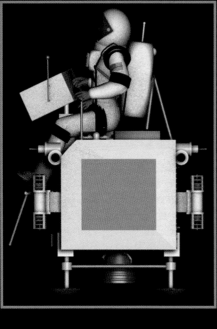
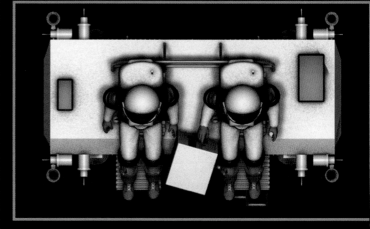

The F2E was modified to also carry three crew (the F3A through F3E). All variants looked basically the same.

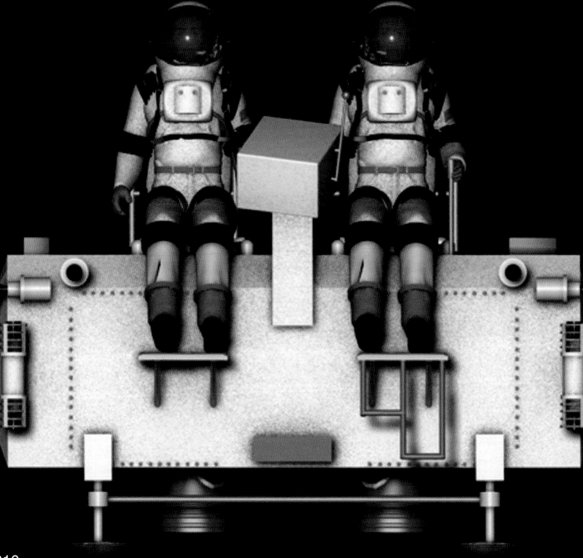

[SHELAB Escape pack 1964]

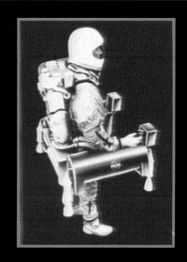

Pictures of the SHELAB rocket pack during testing (above)

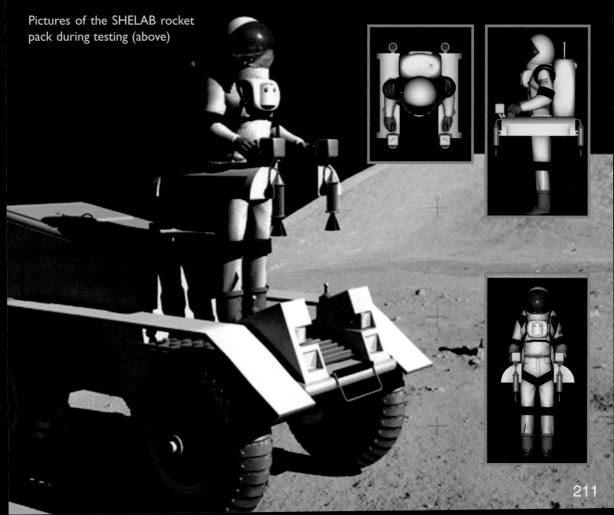

[Bell Flyers 1966]

During the MOBEV studies of 1966 Bell Aerospace Systems looked at a multitude of methods for placing rockets on a small platform. Everything was considered, from small exploration vehicles to complex high-powered escape vehicles capable of allowing a crew to return to lunar orbit. These two pages include some of the rejected configurations. The LFV concept was seriously considered for the Extended LM J-Missions. The geochemist community and NASA's Johnson Space Center were all in favor of using such vehicles for exploration. However, in 1968 Max Faget, who by that time was a very highly respected senior engineer, made the assertion that the proposed LFVs would get progressively heavier as they tried to accomodate safety issues and that training would be even more difficult than it had been with the LLTV. This was accepted as a logical assessment and the LFV was shelved in favor of the rover.

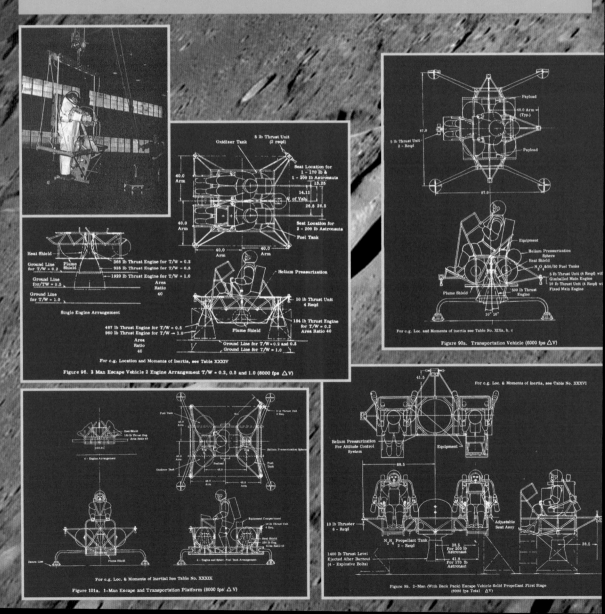

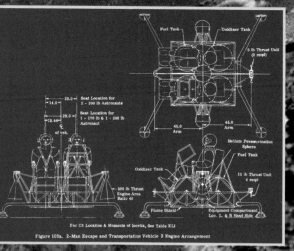

Figure 103a. 2-Man Escape and Transportation Vehicle 2 Engine Arrangement

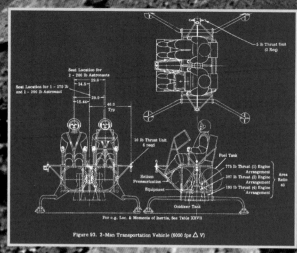

Figure 93. 2-Man Transportation Vehicle (6000 fps ΔV)

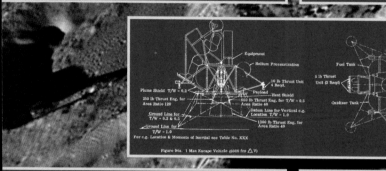

Figure 94a. 1 Man Escape Vehicle (6000 fps ΔV)

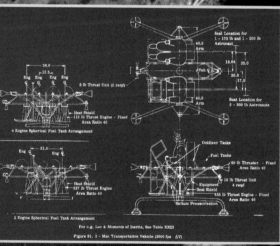

Figure 91. 2 - Man Transportation Vehicle (2000 fps ΔV)

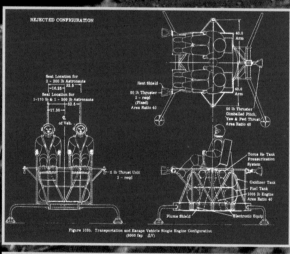

Figure 103b. Transportation and Escape Vehicle Single Engine Configuration (8000 fps ΔV)

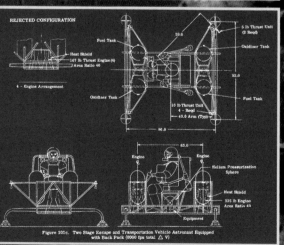

Figure 101c. Two Stage Escape and Transportation Vehicle Astronaut Equipped with Back Pack (8000 fps total ΔV)

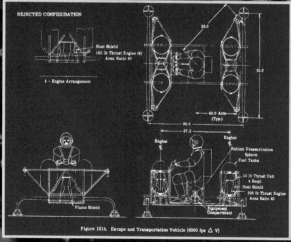

Figure 101b. Escape and Transportation Vehicle (8000 fps ΔV)

[Bell Flyers 1969]

Even long after the MOBEV studies were completed in 1966, Bell Aerospace Systems of Buffalo New York continued to study flying vehicles for use on the lunar surface. Between the original studies of the early sixties, the MOBEV study and the One Man Flying Vehicle report of July 1969, Bell had examined literally dozens of candidate systems. Some of the 1969 designs are shown here.

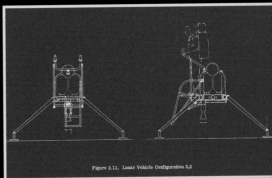
Figure 2.11. Lunar Vehicle Configuration 3.2

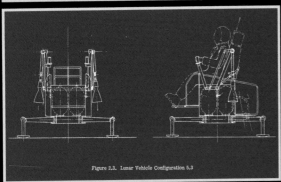
Figure 2.3. Lunar Vehicle Configuration 5.3

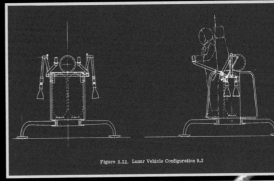
Figure 2.12. Lunar Vehicle Configuration 9.2

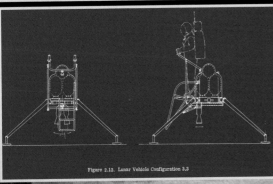
Figure 2.13. Lunar Vehicle Configuration 3.3

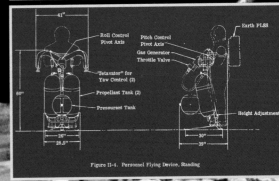
Figure II-4. Personnel Flying Device, Standing

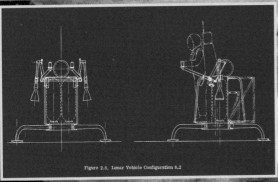
Figure 2.8. Lunar Vehicle Configuration 8.2

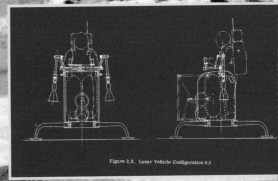
Figure 2.9. Lunar Vehicle Configuration 8.1

[Other Flyers 1966-69]

The NAA flyer (bottom), the MMU (middle right) and Croft's 1969 design (right)

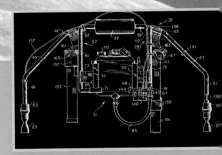

Design work on one-man and two-man flying vehicles was not solely restricted to *Bell Aerospace Systems*. In 1964 *North American Aviation* had also put extensive work into a one-man vehicle that was designed to be a supplement to MOLAB long duration missions. This proposed system would have used a bi-propellant engine and had a range of 16 kilometers. Provisions were made to carry a second astronaut for rescue missions. A simulator for this device was built by the *Space and Information Systems Division* of *North American* and showed that the platform configuration shown here would have been stable and reliable. The vehicle would have been controlled "kinesthetically" in both yaw and roll (which means the pilot threw around his body weight like on a Segway transporter). Pitch control was effected using a secondary set of thrusters.

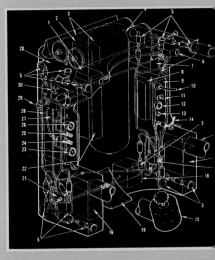

In 1969 design engineer Robert Croft created a lighter version of Moore's original backpack for NASA. Croft's design was specifically designed for lunar use and would not be powerful enough to function on the Earth. Two of them could also be used to move cargo.

Of course the only one-man rocket powered transportation device that actually flew in space was never used on the moon. The *Manned Maneuvering Unit* was built around the same time as these other lunar rocket devices were being studied and was sent into space during the Gemini program. However, due to problems during EVA the astronauts were never able to actually don the back-pack and fly it until many years later, when a modernized version would be used, most famously by Bruce McCandless, during the Space Shuttle program.

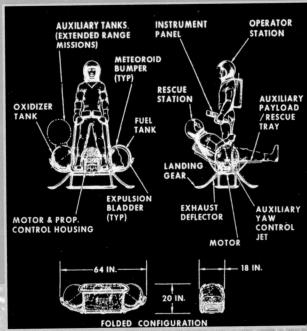

216

[Other Lunar Vehicles 1962-67]

Lunar exploration vehicles were not restricted to rovers and flyers. An assortment of other proposals were entertained in the early 1960's. General Electric's Ralph Mosher worked steadfastly on a human exoskeleton and walking machine. The first iteration from 1965 was known as the Hardiman and, had the computing power been available, it would have allowed an astronaut to walk in a servo-assisted suit which endowed superhuman strength to the wearer. Mosher's research led to a large walking lunar "Pedipulator" and ultimately to the GE Walking Truck. Modern variants of this walking truck are in use today. Funding for Mosher's work came from the DOD and NASA. An earlier version of a walker by Space General (see Mechanix Illustrated April 1962) appeared to have many similar functions to the GE Walking Truck and was also to have been used on the moon. Meanwhile, General Motors considered a corkscrew vehicle or a tracked rover (below) while in 1966, Ford's Aeronutronic proposed a large Lunar Worm that would undulate across the surface. RCA favored either an inflatable plastic ball, with a large solar panel umbrella and counterweights that would roll around the surface, or a variety of different walking machines similar to the Space General walker (see Life Magazine 27/4/62).

Lunar concepts (*clockwise from top left*) GE Walking Truck, GM tracked vehicle, GE Pedipulator, GM Corkscrew, Ford Lunar Worm, GE Hardiman exoskeleton.

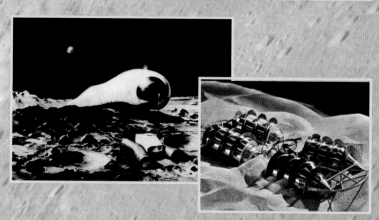

[GM/Boeing LRV 1971]

Conclusion

The extent and scope of the ideas presented here show a diversity of imagination that seems to have been bred out of sheer enthusiasm as well as some not inconsiderable generosity on the part of the American congress and senate. Had Apollo not been dismantled so hastily we might well have seen some of these well-conceived ideas trundling around the lunar surface. But as history played out we would only get to see three lunar modules and three extended lunar modules sent to the Moon. The flights of Apollo 15, 16 and 17 would all carry the brilliantly designed *Boeing/GM* LRV to the lunar surface. It would function almost flawlessly, with the exceptions of a short glitch on Apollo 16 wherein it wouldn't start, and on Apollo 17 where it lost a fender.

The Apollo LRV took many of the best design characteristics of the *Bendix* studies and combined them with expertise from *Goodyear*, *General Motors*, *Grumman* and of course *Boeing*. It was not equipped with a flying rocket back pack like the SHELAB rover, and it couldn't support the crew in the case of an emergency but it did perform its prescribed function brilliantly and many people think that the three abandoned at Hadley, Descartes and Taurus Littrow

The extraordinary GM/Boeing LRV on the Moon

BOEING/GM LRV

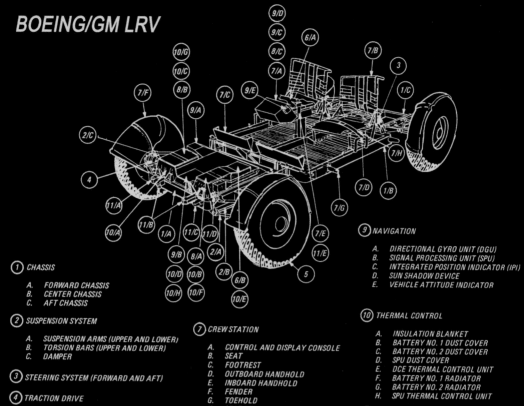

1. CHASSIS
   A. FORWARD CHASSIS
   B. CENTER CHASSIS
   C. AFT CHASSIS

2. SUSPENSION SYSTEM
   A. SUSPENSION ARMS (UPPER AND LOWER)
   B. TORSION BARS (UPPER AND LOWER)
   C. DAMPER

3. STEERING SYSTEM (FORWARD AND AFT)

4. TRACTION DRIVE

5. WHEEL

6. DRIVE CONTROL
   A. HAND CONTROLLER
   B. DRIVE CONTROL ELECTRONICS (DCE)

7. CREW STATION
   A. CONTROL AND DISPLAY CONSOLE
   B. SEAT
   C. FOOTREST
   D. OUTBOARD HANDHOLD
   E. INBOARD HANDHOLD
   F. FENDER
   G. TOEHOLD
   H. SEAT BELT

8. POWER SYSTEM
   A. BATTERY NO. 1
   B. BATTERY NO. 2
   C. INSTRUMENTATION

9. NAVIGATION
   A. DIRECTIONAL GYRO UNIT (DGU)
   B. SIGNAL PROCESSING UNIT (SPU)
   C. INTEGRATED POSITION INDICATOR (IPI)
   D. SUN SHADOW DEVICE
   E. VEHICLE ATTITUDE INDICATOR

10. THERMAL CONTROL
    A. INSULATION BLANKET
    B. BATTERY NO. 1 DUST COVER
    C. BATTERY NO. 2 DUST COVER
    D. SPU DUST COVER
    E. DCE THERMAL CONTROL UNIT
    F. BATTERY NO. 1 RADIATOR
    G. BATTERY NO. 2 RADIATOR
    H. SPU THERMAL CONTROL UNIT

11. PAYLOAD INTERFACE
    A. TV CAMERA RECEPTACLE
    B. LCRU RECEPTACLE
    C. HIGH GAIN ANTENNA RECEPTACLE
    D. AUXILIARY CONNECTOR
    E. LOW GAIN ANTENNA RECEPTACLE

Distances traveled	
Apollo 15	17.3 miles
Apollo 16	16.7 miles
Apollo 17	22.2 miles

Weight:	462 lbs (77 lbs on the moon)
Length:	122 inches
Width:	72 inches
Power supply:	Two 36 volt silver-zinc batteries
Range:	Up to cumulative distance of 57 miles
Drive:	Four ¼-hp DC motors
Wheel base:	90 inches
Clearance:	14 inches
Turning radius:	120 inches
Max Speed:	8.7 mph

GM/Boeing LRV during stowage procedures before flight (below)

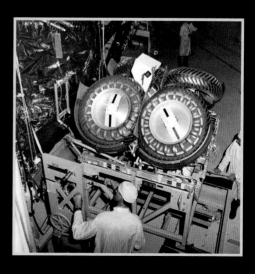

LUNAR ROVING VEHICLE Initial Deployment Sequence

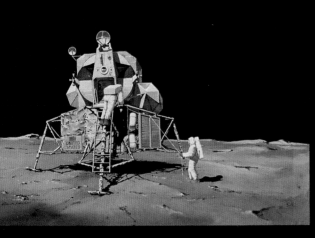
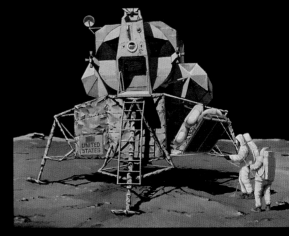
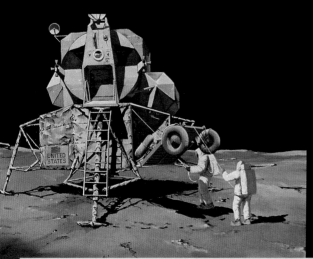
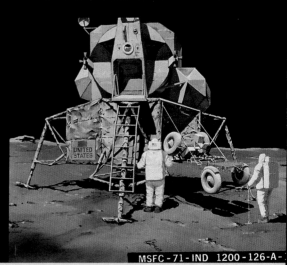

MSFC - 71 - IND 1200 - 126 - A -

may yet still work if someone were to carry some new batteries to the Moon and power them up. The USGS *Grover* is still on display at the USGS *Shoemaker Center for Astrogeology* in Flagstaff Arizona, while *Boeing* have at least one mock-up of the LRV at the *Museum of Flight* in Seattle Washington. A full flight-ready version of the *Boeing*/GM LRV can be seen at the *National Air & Space Museum* in Washington DC. The USGS MGL MOLAB ended up as an exhibit at the *US Space and Rocket Center* in Huntsville while the prototype of the *Grumman MOLAB* is now on display at the *Cradle of Aviation Museum* in Long Island, New York.

Meanwhile the final *Bell* rocket back-pack became a Hollywood celebrity when *Bell* test-pilot William Suitor doubled for *James Bond* and used it to escape the villains in the movie *Thunderball*. An entire cottage industry has grown around the *Bell* rocket packs. Amateur enthusiasts, often working in consort with the original people from *Bell*, have built a vast array of personal flying devices over the last forty years. Sporadic competitions are held at the *Bell Museum* in Niagara Falls, New York, to see who has built the best flyer. One of the original vehicles is on display at that same location.

The conception and construction of the world's first manned lunar landing vehicle is a story of genius and inventiveness rarely seen. Rising above a fierce field of competitors the team at *Grumman* prevailed in the bidding war and built a vehicle of remarkable capability, resilience and reliability. The pictures you have seen in this book demonstrate just how versatile and flexible the LM could have been. Sadly it was only a fantasy, and several fully constructed lunar modules were actually misplaced for many years before finding a home in museums across America. You can see this remarkable vehicle in all of its glory at the *Kennedy Space*

LUNAR ROVING VEHICLE   Final Deployment Sequence

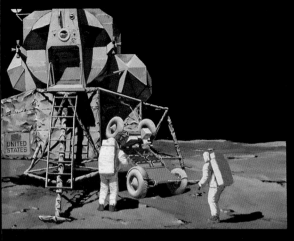
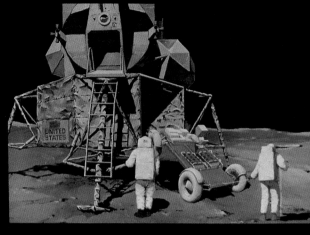

The GM/Boeing LRV steering and control console (above)

Centre in Florida, at the *National Air and Space Museum* (NASM) in Washington DC and at the *Cradle of Aviation* Museum in Long Island, New York. Full size mock-ups are on display at the *Adler* in Chicago, outside the gates at the NASA *Stennis Space Center* in Mississippi and at the *US Space & Rocket Center* in Huntsville Alabama. Some of the early desk models of the LEM are also on display at the NASM, including the five-legged version from 1962. Diagrams from *Project Horizon* can be seen scattered across Wernher von Braun's desk in a display at the *Stennis Space Center* in Mississippi.

Sadly, many of the documents, models and photographs have been lost or misplaced over the years. There are no monuments to Rosen's giant *NOVA* or models of Martin's remarkable inflatable landing gear. The full-size MOLAB built in Huntsville and Chrysler's nuclear tricycle seem to have been lost or perhaps dismantled, and the desk models pictured in this book have all vanished. No one seems to know what happened to LM TM-1 or the metal M-5 mock-up or the *Bendix* RIBE test article. All of these fragments of American history are now scattered like pages torn from an old scrapbook.

INDEX

2-Man Apollo 56, 61
3-man LM 118, 120-121
4 man intermediate recon rover 103
6-Man Apollo 122-123

A

Aberdeen Proving Ground 133
ADAM 119-120, 123
ADAM/SM 120, 123
Adler Planetarium 221
Advanced Mercury 36
Aerojet 12-15
Aerojet Moonmobile 13
Allen, John Jr 70
ALM Shelter 78
ALM Taxi 78
ALSS 86, 189, 198
Ames Research Centre 27
Apollo 14 112
Apollo 20 138
Apollo 2-man Direct 54
Apollo 5 76
Apollo 5 LM 76
Apollo Applications Program 24, 78, 131, 135, 154
Apollo CSM 27, 41, 49, 114-115, 120, 144, 181
Apollo Multipurpose Mission Module 160
Augmented LM 78
Augmented LMS 78
Augmented LMT 78, 88
Augmented LMTS 78
Augmented LPM 78
Avro Arrow 36, 42

B

B1 reentry vehicle 27
Bell Aerospace 25, 28, 68, 70, 114, 124, 126, 144, 189, 198, 200-209, 212, 214, 216, 220
Bell flyers 212, 214
Bell Hopper 189, 198
Bell LFV 203
Bell Museum 220
Bell rocket pack 201-202
Bendix 24, 64, 112, 124, 126-129, 131-133, 138-139, 143, 150, 154, 157-160, 162-163, 171-172, 186, 188, 198, 201, 218, 221
Bendix Dual LRV 143
Bendix LEM 64
Bendix Mobility Test Article BX-1 132
Bendix MOLEM 158
Bendix SLRV Surveyor 129
Bendix two wheel 112
BIS 4-5, 34, 38
BIS lunar lander 4
Boeing 1-man hopper 106
Boeing 24, 25, 51-52, 72, 86, 99-108, 110-111, 126-128, 133, 135-136, 155-157, 160-161, 181-182, 198, 218-221
Boeing 2-man exploration vehicle 105
Boeing CAN 160
Boeing LEM 51
Boeing LRV 86, 155, 218, 221
Boeing V-1 AMF 110
Boeing/GM LRV 218-220
Boeing/GM MOLAB 181
Boeing/GMDRL LSSM 135
Bonestell, Chesley 6, 63

British Interplanetary Society 4, 34

C

C-1 reentry vehicle 27
C-2 reentry vehicle 27
CF-105 Arrow 42
Chamberlin, Jim 36, 38, 42
Chrysler Corporation 98, 113, 221
Chrysler LRV 98
Chrysler Lunar Freight Vehicle 98
Cinder Lake 128, 130, 139-140, 153-155
Clarke, Arthur C. 4-5
CM logistics vehicle 48
CM Shelter 85, 159
Collier's magazine 6
Collins, William 208
Conquest of the Moon 6
Conrad, Pete 74
Convair 25-28
Convair/Avco 25
Cornell/Bell/Raytheon 25
Cradle of Aviation Museum 220-221
Croft, Robert 216

D

D-2 27, 48
Design 378B 90
Direct sidelander 18
Douglas 25, 146
Dryden Research Center 68
Dyna-Soar 32

E

Earth Orbit Rendezvous 20, 28, 53
Edwards, Happian 4
elastic conoid 127
ELM Taxi 78
ELS (Early Lunar Shelter) 88-89, 124-125
EOR 20, 32, 38, 40, 53
Explorer 154, 161, 163, 170, 175
Extended ELS 124-125
Extended LM 76, 78, 114, 122, 137
Extended Lunar Operations Study 67

F

F-1 8, 10, 20, 22, 38, 49
F1A 204
F1B 203-205
F1C 206
F2B 203, 207-208
F2E 203, 209-210
Faget, Max 38, 40, 54
FD LEM 44, 45
FFC flat faced cone reentry vehicle 27
Fleming Ad Hoc Task Group 16
Fleming, William 16, 19
Flight Hover Indexer 107
FLOX 123
fluorine 123
flying saucers 38
Ford Aeronutronic 218
Further Developments LEM 44

G

Gainor, Chris 42
Garrett AiResearch 88
GE D-2 27, 48
Gemini 36, 38, 54-57, 80, 216
General Dynamics/Convair 26, 27, 48

General Electric 25, 27-29, 48, 124, 127, 218
General Electric/Bell 25
General Motors Defense Research Laboratory 108
General Motors/Boeing 24, 126, 188
Gernsback, Hugo 198
Gilruth, Robert 61, 74
Glenn, John 47
GM/Boeing 1-G Trainer 156
GM-1 133-134
Goodyear 24, 25, 84, 124, 127, 218
Grover 154-156, 220
Grumman 25, 56, 58, 60-64, 66-68, 72-74, 78-79, 88, 90, 94, 114, 121, 127-128, 130-131, 137-142, 146, 150-151, 157, 164, 198, 218-220
Grumman Dual LRV 137-138
Grumman inflatable LRV 142
Grumman LEM 58, 61-63, 66, 72, 74, 78-79
Grumman LM 94, 114
Grumman LRV 139-140
Grumman MOLAB 150-151, 220
Grumman MOLEM 157
Grumman Rover 88, 137, 141
Grumman space-bike 76
Grumman/ITT 25
Guardite 25

H

Hardiman 218
hard-shell suit 15
Hayes International Corporation 86, 189
Hazard, Allyn B. 14
Heaton, Col D.H. 20
Honzik, George 67
Houbolt, John 34
Huntsville 2, 9, 16, 20, 25, 38, 40, 42, 46, 53, 86, 124, 132, 160, 220-221

I

ILMT 78, 82, 85
INT-21 118-120
Irwin, Jim 155

J

J-2 10, 22, 49, 146-147
J2-S Saturn V 122
Johnson Space Center 70
JPL 8, 14, 25, 162-164

K

Kehlet, Alan 20, 29
Kennedy Space Centre 220
Kennedy, President John 10, 26
Koelle, Heinz-Hermann 8, 25

L

L-I 145
L-II 145
L-III 145
L-1 reentry vehicle 27
L-2C reentry vehicle 27
L-20 28, 30-31
L3A reentry vehicle 27
L4 reentry vehicle 27
L7 27-28
L8 reentry vehicle 27
Landing Training Vehicles 70
Langley Research Center 25, 27, 48

LASS 108, 146-148, 163, 193
LASSO 122, 148
LBV 67
LEM 40, 42, 44, 46, 49, 51, 58, 60-64, 66-68, 72-76, 78-79, 90, 135, 221
Lenticular 26, 28-29
LESA 99-101, 105-106, 124-125, 145
Levin, Kenneth 70
LIFE magazine 14
lifting bodies 38
LM 40
LRF 68-69
LRV 68, 70-72
LS 119
LTV 70
LV 3, 67, 99, 104, 124-125, 131, 145-147, 149, 163, 179, 185, 193
LM controls 96
LM Lab 90
LM Shelter 78-79, 80, 88, 124, 141, 157, 163
LM Taxi 78-79, 114, 150
LM TM-1 72, 221
LM Truck 78-79, 82, 84, 86, 88, 124, 131, 145, 150, 152, 157, 159, 163, 181, 188-189
LM Truck Shelter 78, 82, 84
LM Truck Power Plants 124
LM Truck U (uprated) 124
LM/ATM 93
LM/Stellar ATM 93
MDV 153
Local Scientific Survey Module 170
Lockheed 25, 67, 99, 121, 123-124, 126
Lovell, Jim 74
LPM 78, 80-81, 122
LSSM 78, 126-127, 131, 133, 135-136, 154, 157, 159-164, 168, 170-175, 177
LSV 108
LTS 119
LTV 67
Lunar Application Spent Stage Orbital 122
Lunar Applications of a Spent S-IVB/IU Stage 146, 163
Lunar Direct 10, 16-18, 20, 28, 38, 40, 43, 48, 51, 56, 61-62, 119-120
Lunar Excursion Module 40, 49-50
Lunar Exploration Systems for Apollo 99
Lunar Gemini 54, 56-57
Lunar Habitat 124, 146
Lunar Landing Research Facility 68
Lunar Landing Research Vehicle 68, 71
Lunar Landing Stage 119
Lunar materials handling vehicle 104
Lunar Mission Development Vehicle 153
Lunar Module 40, 43, 72, 76, 83, 96, 121, 199, 201
Lunar Payload Module 78, 80
Lunar surface rendezvous 25
Lunar Surface Vehicle 108
Lunar Takeoff Stage 119
Lunar Worm 218
LUNEX 8, 32-33
Lunokhod 24

M1 reentry vehicle 12, 20. 26, 31, 48
M2 reentry vehicle 27
M5 74, 86, 221
MALLIR 34-35
Manned Lunar Auxiliary Vehicle 113
Manned Maneuvering Unit 216
Markow, Edward 127
Marquardt Space Sled 76-77

Mars Rovers 127, 164
Marshall Space Flight Centre 53, 99, 189
Martin 20, 25, 28-31, 221
Martin Corporation 20, 48, 193
Martin Model 410 27, 30-31, 48
Mason, Major Matt 14
Mattel corporation 14
Matzenauer, James 85
Maximum Concept Cargo Hauler 109
Maynard, Owen 42, 45, 74
McDivitt, Jim 60
McDonnell 25, 54, 61
Mercury 11, 20, 31-32, 36, 38, 40, 50, 54
MET 112
methane 123
Michael, William H. Jr 34
MIMOSA 124
MISDAS 48
Mk1 Moonmobile 13
MLMT 78, 83
MOBEV 126, 157, 161, 163, 169, 171, 175, 181, 186, 189, 193, 201, 205, 212, 214
Mobile Equipment Transporter 112
MOBEX Mobile Excursion Laboratory 78, 184-187
Mobile Geologic Laboratory 153
Mobility Test Article 132-134
MOCAN 157, 160-161, 179
MOCOM 126, 157, 159-161, 179
Modified Mercury 28, 31, 48
Modular Lunar Service Truck 83
MOLAB 3, 24, 78, 127, 132-134, 150-153, 160-164, 175, 179, 181-182, 184, 189, 194, 196, 198, 203, 216, 220-221
MOLAB 944-001 181
MOLEM 126, 157-158, 160-161, 179
Moon Car 127
Moonsuit 14-15
Moore, Wendell 198, 203-205
Mosher, Ralph 218
MTA GM-1 133
Museum of Flight 220

N

NAA Flyer 216
NAA Winged Apollo 144
NACA 28
National Air & Space Museum 220
North American 25, 28, 38, 40, 48, 61, 74, 114, 118, 120-121, 123, 144, 149, 216
North American Rockwell 85, 114, 118, 120-121, 144
Northrop 146, 188-189
Northrop WBS MOLAB 188
Nova 10-12, 16, 19-20, 26, 38, 40, 221
Nuclear LESA 124
nuclear LV-7 third stage 118

O

Oberth, Hermann 127
one-man escape vehicle 200

P

paraglider landing system 56
Pedipulator 218
Petersen, Bob 127
plywood 130
POGO 203-204, 206
Postle, Robert 28
President's Scientific Advisory Committee 54

Project Horizon 8, 20, 221
Prospector Program 24
PSAC 54

R

R0AB 162-163, 190-192, 195
R0AE 161-162, 164-165
R0BB 162-163, 194, 196
R0BE 161-162, 167, 169
R0CB 162-163, 197
R0CE 161-162, 164, 166
R0DE 161, 170
R-1 reentry vehicle 27
R1A(1)E 161-162, 169
R1AB 162, 190
R1AE 161-162, 167-169
R1B(1)E 161-162, 174, 190, 195
R1BB 162, 191-192
R1BE 131, 161-163, 170-172, 221
R1CB 162-163, 193
R1CE 161-162, 175-176, 178, 191
R1DE 161-162, 164, 174-178, 192
R2 reentry vehicle 27
R2AB 162
R2AE 161, 170-171, 173
R2BB 162, 194
R2BE 161, 179-180
R2C(1)E 161-163, 179-180, 194, 196
R2C(2)E 161, 179
R2C(3)E 161, 179
R2C(4)E 161, 179
R2C(5)E 161, 179, 181
R2DE 161, 179-180
R-3 reentry vehicle 27, 29
R3AE 161-162, 184, 187
R3BE 161-163, 184, 187, 197
R3CE 161-162, 179-180
R3DE 161-162, 175, 185, 187
R4AE 161, 184, 186
RCA 218
Redstone Arsenal 20
Republic 15, 25
Republic Aviation 15
Rescue LM 92
RL-10 51, 123, 145
rocket pack 201-202, 211
Rocketdyne 8, 22
Rockwell Apollo 28
Rosen, Milton 8, 38
Rosen/Schwenk 10, 12
Ross, H.E. 4, 34

S

Saturn 9-10, 20, 22, 23, 38, 40, 49, 54-57, 61-62, 64, 76, 86, 90, 98-99, 103, 108-109, 114, 116, 118-120, 122, 124, 146, 183
Saturn C-1 24, 38
Saturn C-2 27, 38
Saturn C-3 9, 20, 38
Saturn C-5 20
Saturn IB 76, 90
Saturn Lunar Adapter 114
Saturn V 21-23, 49, 54-57, 61-62, 78, 90, 98-99, 103, 109, 114, 116, 118-120, 122, 124, 146, 183
SB-625 LLS 145
SB-910 145
SB-908 145
SB-911 145
SB-810 LLS 145
SB-529 LLS 145

SB-809 LLS 145
SB-713 145
Schmitt, Harrison 140
Schueller, Otto 15
Schwenk, Carl 8
Schwinn Bicycle Company 24
Scott, Dave 155
SEC 14, 34
service module injection stage 119
Service Module Logistic Vehicle 149
Shea, Joe 47, 49-50, 54, 61, 74
SHELAB 3, 86-87, 211, 218
SHELAB Escape Pack 211
Shepard, Alan 74
Shepard, Leonard 14
Shoaf, Harry C. 37
Shoemaker Center for Astrogeology 220
SIC 22
S-II 22, 49, 122, 147
S-IVB 22, 49, 122, 146-147, 163
Skylab 118
SLA 82, 86, 114, 118, 124, 160, 181, 183
SLA mini-base 114, 118
SLAMB 114-120
Slayton, Deke 74
SLRV 128-129, 133, 161-165
SLRV Surveyor 128-129
SMIS 116, 119-120, 122
Smith, R.A. 4-5
Smyth, Robert 74
SNAP 19 138-139
SNAP 27 88, 164
SNAP 29 164
SNAP 9A 113

SNAP-11 RTG 128
Society of Automotive Engineers 98
Soyuz 27
Space Craft Inc 25
Space News 40, 45, 60, 69, 203
Space Systems Division 32
Space Task Group 36, 38, 42
Space Technology Laboratories Inc 61
Spacecraft LM Adapter 82
Space-General Corporation 14, 218
Specialized Construction Vehicle 101
Specified LSSM 131, 136
Sputnik 8, 32
Star Trek 29
Stay Time Extension Module 84
STEM 78, 84, 124
Stennis Space Center 221
STL Apollo 61
STL Direct 61
Stuhlinger, Ernst 8, 47
Suitor, William 220
Surveyor Lunar Roving Vehicles 129

T

10th International Astronautical Congress 10
Thiokol 198
TRW 145
two-man intermediate recon rover 102

U

Unitized Local Rover 100, 102
US Geological Survey 128
US Space and Rocket Center 220

USGS 2, 127-128, 130, 153-155, 220
USGS cloth LM 130
USGS Geologic Rover 155
USGS LMs 130

V

V-1 GMB 111
Vehicle Assembly Building 23
von Braun, Wernher 6, 8-9, 47, 132, 221
Von Braun's 1952 lander 7
Vought 25

W

W-1 reentry vehicle 27
Walker, Joe 70
Walking Beam Suspension 163, 188
Walking Truck 218
Webb, James 27, 53
Westinghouse 183
White, Ed 68
Wiesner, Jerome 53
Williams, Walter 38
wire wheel 127

X

X-20 29, 32

Y

Young, John 130, 154
Yuma Proving Ground